Big Data on Kubernetes

A practical guide to building efficient and
scalable data solutions

Neylson Crepalde

Big Data on Kubernetes

Group Product Manager: Apeksha Shetty

Publishing Product Manager: Apeksha Shetty

Book Project Manager: Aparna Ravikumar Nair

Senior Editor: Sushma Reddy

Technical Editor: Kavyashree K S

Copy Editor: Safis Editing

Proofreader: Sushma Reddy

Indexer: Subalakshmi Govindhan

Production Designer: Gokul Raj S T

DevRel Marketing Executive: Nivedita Singh

First published: July 2024

Production reference: 1210624

Published by Packt Publishing Ltd.

Grosvenor House

11 St Paul's Square

Birmingham

B3 1RB, UK

ISBN: 978-1-83546-214-0

www.packtpub.com

To my wife, Sarah, and my son, Joao Pedro, for their love and support. To Silvio Salej Higgins for being a great mentor.

– Neylson Crepalde

Contributors

About the author

Neylson Crepalde is a generative AI strategist at **Amazon Web Services** (**AWS**). Before this, Neylson was chief technology officer at A3Data, a consulting business focused on data, analytics, and artificial intelligence. In his time as CTO, he worked with the company's tech team to build a Big Data architecture on top of Kubernetes that inspired the writing of this book. Neylson holds a PhD in economic sociology, and he was a visiting scholar at the *Centre des Sociologies des Organisations* at Sciences Po, Paris. Neylson is also a frequent guest speaker at conferences and has taught MBA programs for more than 10 years.

I want to thank all the people who have worked with me in the development of this great architecture, especially Mayla Teixeira and Marcus Oliveira for their outstanding contributions.

About the reviewer

Thariq Mahmood has 16 years of experience in data technology and possesses a strong skillset in Kubernetes, Big Data, data engineering, and DevOps on the public cloud, the private cloud, and on-premise environments. He has expertise in data warehousing, data modeling, and data security. He actively contributes to projects on Git and has experience setting up batch and streaming pipelines for various production environments, using Databricks, Hadoop, Spark, Flink, and other cloud-native tools from AWS, Azure, and GCP. Also, he implemented MLOps and DevSecOps in numerous projects. He currently works on helping organizations optimize their Big Data infrastructure costs and implementing data-lake and one-lake architectures within Kubernetes.

Table of Contents

Part 2: Big Data Stack

4

The Modern Data Stack   61

5

6

7

Part 3: Connecting It All Together

11

Generative AI on Kubernetes   219

12

Where to Go from Here   253

Index   263

Other Books You May Enjoy   272

Preface

In today's data-driven world, the ability to process and analyze vast amounts of data has become a critical competitive advantage for businesses across industries. Big data technologies have emerged as powerful tools to handle the ever-increasing volume, velocity, and variety of data, enabling organizations to extract valuable insights and drive informed decision-making. However, managing and scaling these technologies can be a daunting task, often requiring significant infrastructure and operational overhead.

Enter Kubernetes, the open source container orchestration platform that has revolutionized the way we deploy and manage applications. By providing a standardized and automated approach to container management, Kubernetes has simplified the deployment and scaling of complex applications, including big data workloads. This book aims to bridge the gap between these two powerful technologies, guiding you through the process of implementing a robust and scalable big data architecture on Kubernetes.

Throughout the chapters, you will embark on a comprehensive journey, starting with the fundamentals of containers and Kubernetes architecture. You will learn how to build and deploy Docker images, understand the core components of Kubernetes, and gain hands-on experience in setting up local and cloud-based Kubernetes clusters. This solid foundation will prepare you for the subsequent chapters, where you will dive into the world of the modern data stack.

The book will introduce you to the most widely adopted tools in the big data ecosystem, such as Apache Spark for data processing, Apache Airflow for pipeline orchestration, and Apache Kafka for real-time data ingestion. You will not only learn the theoretical concepts behind these technologies but also gain practical experience in implementing them on Kubernetes. Through a series of hands-on exercises and projects, you will develop a deep understanding of how to build and deploy data pipelines, process large datasets, and orchestrate complex workflows on a Kubernetes cluster.

As the book progresses, you will explore advanced topics such as deploying a data consumption layer with tools such as Trino and Elasticsearch and integrating generative AI workloads using Amazon Bedrock. These topics will equip you with the knowledge and skills necessary to build and maintain a robust and scalable big data architecture on Kubernetes, ensuring efficient data processing, analysis, and analytics application deployment.

By the end of this book, you will have gained a comprehensive understanding of the synergy between big data and Kubernetes, enabling you to leverage the power of these technologies to drive innovation and business growth. Whether you are a data engineer, a DevOps professional, or a technology enthusiast, this book will provide you with the practical knowledge and hands-on experience needed to successfully implement and manage big data workloads on Kubernetes.

Who this book is for

If you are a data engineer, a cloud architect, a DevOps professional, a data or science manager, or a technology enthusiast, this book is for you. You should have a basic background in Python and SQL programming, and basic knowledge of Apache Spark, Apache Kafka, and Apache Airflow. A basic understanding of Docker and Git will also be helpful.

What this book covers

Chapter 1, *Getting Started with Containers*, embarks on a journey to understand containers and Docker, the foundational technologies for modern application deployment. You'll learn how to install Docker and run your first container image, experiencing the power of containerization firsthand. Additionally, you'll dive into the intricacies of Dockerfiles, mastering the art of crafting concise and functional container images. Through practical examples, including the construction of a simple API and a data processing job with Python, you'll grasp the nuances of containerizing services and jobs. By the end of this chapter, you'll have the opportunity to solidify your newfound knowledge by building your own job and API, laying the groundwork for a portfolio of practical container-based applications.

Chapter 2, *Kubernetes Architecture*, introduces you to the core components that make up the Kubernetes architecture. You will learn about the control plane components such as the API server, etcd, scheduler, and controller manager, as well as the worker node components such as kubelet, kube-proxy, and container runtime. The chapter will explain the roles and responsibilities of each component, and how they interact with each other to ensure the smooth operation of a Kubernetes cluster. Additionally, you will gain an understanding of the key concepts in Kubernetes, including pods, deployments, services, jobs, stateful sets, persistent volumes, ConfigMaps, and secrets. By the end of this chapter, you will have a solid foundation in the architecture and core concepts of Kubernetes, preparing you for hands-on experience in the subsequent chapters.

Chapter 3, *Kubernetes – Hands On*, guides you through the process of deploying a local Kubernetes cluster using kind, and a cloud-based cluster on AWS using Amazon EKS. You will learn the minimal AWS account configuration required to successfully deploy an EKS cluster. After setting up the clusters, you will have the opportunity to choose between deploying your applications on the local or cloud environment. Regardless of your choice, you will retake the API and data processing jobs developed in *Chapter 1* and deploy them to Kubernetes. This hands-on experience will solidify your understanding of Kubernetes concepts and prepare you for more advanced topics in the following chapters.

Chapter 4, *The Modern Data Stack*, introduces you to the most well-known data architecture designs, with a focus on the "lambda" architecture. You will learn about the tools that make up the modern data stack, which is a set of technologies used to implement a data lake(house) architecture. Among these tools are Apache Spark for data processing, Apache Airflow for data pipeline orchestration, and Apache Kafka for real-time event streaming and data ingestion. This chapter will provide a conceptual introduction to these tools and how they work together to build the core technology assets of a data lake(house) architecture.

Chapter 5, Big Data Processing with Apache Spark, introduces you to Apache Spark, one of the most popular tools for big data processing. You will understand the core components of a Spark program, how it scales and handles distributed processing, and best practices for working with Spark. You will implement simple data processing tasks using both the DataFrames API and the Spark SQL API, leveraging Python to interact with Spark. The chapter will guide you through installing Spark locally for testing purposes, enabling you to gain hands-on experience with this powerful tool before deploying it on a larger scale.

Chapter 6, Apache Airflow for Building Pipelines, introduces you to Apache Airflow, a widely adopted open source tool for data pipeline orchestration. You will learn how to install Airflow using Docker and Astro CLI, making the setup process straightforward. The chapter will familiarize you with Airflow's core features and the most commonly used operators for data engineering tasks. Additionally, you will gain insights into best practices for building resilient and efficient data pipelines that leverage Airflow's capabilities to the fullest. By the end of this chapter, you will have a solid understanding of how to orchestrate complex data workflows using Airflow, a crucial skill for any data engineer or data architect working with big data on Kubernetes.

Chapter 7, Apache Kafka for Real-Time Events and Data Ingestion, introduces you to Apache Kafka, a distributed event streaming platform that is widely used for building real-time data pipelines and streaming applications. You will understand Kafka's architecture and how it scales while being resilient, enabling it to handle high volumes of real-time data with low latency. You will learn about Kafka's distributed topics design, which underpins its robust performance for real-time events. The chapter will guide you through running Kafka locally with Docker and implementing basic reading and writing operations on topics. Additionally, you will explore different strategies for data replication and topic distribution, ensuring you can design and implement efficient and reliable Kafka clusters.

Chapter 8, Deploying the Big Data Stack on Kubernetes, guides you through the process of deploying the big data tools you learned about in the previous chapters on a Kubernetes cluster. You will start by building bash scripts to deploy the Spark operator and run `SparkApplications` on Kubernetes. Next, you will deploy Apache Airflow to Kubernetes, enabling you to orchestrate data pipelines within the cluster. Additionally, you will deploy Apache Kafka on Kubernetes using both the ephemeral cluster and JBOD techniques. The Kafka Connect cluster will also be deployed, along with connectors to migrate data from SQL databases to persistent object storage. By the end of this chapter, you will have a fully functional big data stack running on Kubernetes, ready for further exploration and development.

Chapter 9, Data Consumption Layer, guides you through the process of securely making data available for business analysts in a big data architecture deployed on Kubernetes. You will start by gaining an overview of working on a modern approach using a "data lake engine" instead of a data warehouse. In this chapter, you will become familiar with Trino for data consumption directly from a data lake through Kubernetes. You will understand how a data lake engine works, deploy it into Kubernetes, and monitor query execution and history. Additionally, for real-time data, you will get familiar with Elasticsearch and Kibana for data consumption. You will deploy these tools, and learn how to index data in them and how to build a simple data visualization with Kibana.

Chapter 10, *Building a Big Data Pipeline in Kubernetes*, guides you through the process of deploying and orchestrating two complete data pipelines, one for batch processing and another for real-time processing, on a Kubernetes cluster. You will connect all the tools you've learned about throughout the book, such as Apache Spark, Apache Airflow, Apache Kafka, and Trino, to build a single, complex solution. You will deploy these tools on Kubernetes, write code for data processing and orchestration, and make the data available for querying through a SQL engine. By the end of this chapter, you will have hands-on experience in building and managing a comprehensive big data pipeline on Kubernetes, integrating various components and technologies into a cohesive and scalable architecture.

Chapter 11, *Generative AI on Kubernetes*, guides you through the process of deploying a generative AI application on Kubernetes using Amazon Bedrock as a service suite for foundational models. You will learn how to connect your application to a knowledge base serving as a **Retrieval-Augmented Generation** (**RAG**) layer, which enhances the AI model's capabilities by providing access to external information sources. Additionally, you will discover how to automate task execution by the AI models with agents, enabling seamless integration of generative AI into your workflows. By the end of this chapter, you will have a solid understanding of how to leverage the power of generative AI on Kubernetes, unlocking new possibilities for personalized customer experiences, intelligent assistants, and automated business analytics.

Chapter 12, *Where to Go from Here*, guides you through the next steps in your journey toward mastering big data and Kubernetes. You will explore crucial concepts and technologies that are essential for building robust and scalable solutions on Kubernetes. This includes monitoring strategies for both Kubernetes and your applications, implementing a service mesh for efficient communication, securing your cluster and applications, enabling automated scalability, embracing GitOps and CI/CD practices for streamlined deployment and management, and Kubernetes cost control. For each topic, you'll receive an overview and recommendations on the technologies to explore further, empowering you to deepen your knowledge and skills in these areas.

To get the most out of this book

Some basics in Python programming knowledge and experience with Spark, Docker, Airflow, Kafka, and Git will help you get the most out of this book.

Software/hardware covered in the book	Operating system requirements
Python>=3.9	Windows, macOS, or Linux
Docker, the latest version available	Linux
Docker Desktop, the latest version available	Windows or macOS
Kubectl	Windows, macOS, or Linux
Awscli	Windows, macOS, or Linux

Software/hardware covered in the book	Operating system requirements
Eksctl	Windows, macOS, or Linux
DBeaver Community Edition	Windows, macOS, or Linux

All guidance needed for software installation will be provided in each chapter.

If you are using the digital version of this book, we advise you to type the code yourself or access the code from the book's GitHub repository (a link is available in the next section). Doing so will help you avoid any potential errors related to the copying and pasting of code.

Download the example code files

You can download the example code files for this book from GitHub at `https://github.com/PacktPublishing/Bigdata-on-Kubernetes`. If there's an update to the code, it will be updated in the GitHub repository.

We also have other code bundles from our rich catalog of books and videos available at `https://github.com/PacktPublishing/`. Check them out!

Conventions used

There are a number of text conventions used throughout this book.

`Code in text`: Indicates code words in text, database table names, folder names, filenames, file extensions, pathnames, dummy URLs, user input, and Twitter handles. Here is an example: "This command will pull the `hello-world` image from the Docker Hub public repository and run the application in it. "

A block of code is set as follows:

```
import pandas as pd

url = 'https://raw.githubusercontent.com/jbrownlee/Datasets/
master/pima-indians-diabetes.data.csv'

df = pd.read_csv(url, header=None)

df["newcolumn"] = df[5].apply(lambda x: x*2)

print(df.columns)
print(df.head())
print(df.shape)
```

Any command-line input or output is written as follows:

```
$ sudo apt install docker.io
```

This is how the filename above the code snippet will look:

Cjava.py

Bold: Indicates a new term, an important word, or words that you see onscreen. For instance, words in menus or dialog boxes appear in **bold**. Here is an example: "You should ensure that the **Use WSL 2 instead of Hyper-V** option is selected on the **Configuration** page."

> **Tips or important notes**
> Appear like this.

Get in touch

Feedback from our readers is always welcome.

General feedback: If you have questions about any aspect of this book, email us at customercare@packtpub.com and mention the book title in the subject of your message.

Errata: Although we have taken every care to ensure the accuracy of our content, mistakes do happen. If you have found a mistake in this book, we would be grateful if you would report this to us. Please visit www.packtpub.com/support/errata and fill in the form.

Piracy: If you come across any illegal copies of our works in any form on the internet, we would be grateful if you would provide us with the location address or website name. Please contact us at copyright@packt.com with a link to the material.

If you are interested in becoming an author: If there is a topic that you have expertise in and you are interested in either writing or contributing to a book, please visit authors.packtpub.com.

Share Your Thoughts

Once you've read *Big Data on Kubernetes*, we'd love to hear your thoughts! Scan the QR code below to go straight to the Amazon review page for this book and share your feedback.

https://packt.link/r/1-835-46214-6

Your review is important to us and the tech community and will help us make sure we're delivering excellent quality content.

Download a free PDF copy of this book

Thanks for purchasing this book!

Do you like to read on the go but are unable to carry your print books everywhere?

Is your eBook purchase not compatible with the device of your choice?

Don't worry, now with every Packt book you get a DRM-free PDF version of that book at no cost.

Read anywhere, any place, on any device. Search, copy, and paste code from your favorite technical books directly into your application.

The perks don't stop there, you can get exclusive access to discounts, newsletters, and great free content in your inbox daily

Follow these simple steps to get the benefits:

1. Scan the QR code or visit the link below

https://packt.link/free-ebook/978-1-83546-214-0

2. Submit your proof of purchase
3. That's it! We'll send your free PDF and other benefits to your email directly

Part 1:
Docker and Kubernetes

In this part, you will learn about the fundamentals of containerization and Kubernetes. You will start by understanding the basics of containers and how to build and run Docker images. This will provide you with a solid foundation for working with containerized applications. Next, you will dive into the Kubernetes architecture, exploring its components, features, and core concepts such as pods, deployments, and services. With this knowledge, you will be well equipped to navigate the Kubernetes ecosystem. Finally, you will get hands-on experience by deploying local and cloud-based Kubernetes clusters and then deploying applications you built earlier onto these clusters.

This part contains the following chapters:

- Chapter 1, Getting Started with Containers
- Chapter 2, Kubernetes Architecture
- Chapter 3, Kubernetes – Hands On

1
Getting Started with Containers

The world is rapidly generating massive amounts of data from a variety of sources – mobile devices, social media, e-commerce transactions, sensors, and more. This data explosion is often referred to as "big data." While big data presents immense opportunities for businesses and organizations to gain valuable insights, it also brings tremendous complexity in how to store, process, analyze, and extract value from huge volumes of diverse data.

This is where Kubernetes comes in. Kubernetes is an open source container orchestration system that helps automate the deployment, scaling, and management of containerized applications. Kubernetes brings important advantages for building big data systems. It provides a standard way to deploy containerized big data applications on any infrastructure. This makes it easy to migrate applications across on-premises servers or cloud providers. It also makes it simple to scale big data applications up or down based on demand. Additional containers can be spun up or shut down automatically based on usage.

Kubernetes helps ensure the high availability of big data applications through features such as self-healing and auto-restarting of failed containers. It also provides a unified way to deploy, monitor, and manage different big data components. This reduces operational complexity compared to managing each system separately.

This book aims to provide you with practical skills for leveraging Kubernetes to build robust and scalable big data pipelines. You will learn how to containerize and deploy popular big data tools such as Spark, Kafka, Airflow, and more on Kubernetes. The book covers architectural best practices and hands-on examples for building batch and real-time data pipelines.

By the end, you will gain an end-to-end view of running big data workloads on Kubernetes and be equipped to build efficient data platforms that power analytics and artificial intelligence applications. The knowledge you will gain will be of immense value whether you are a data engineer, data scientist, DevOps engineer, or technology leader driving digital transformation in your organization.

The foundation of Kubernetes is containers. Containers are one of the most used technologies in data engineering today. They allow engineers to package software into standardized units for development, shipment, and deployment. By the end of this chapter, you will understand the basics of containers and be able to build and run your own containers using Docker.

In this chapter, we will cover what containers are, why they are useful, and how to create and run containers on your local machine using Docker. Containers solve many problems that developers face when moving applications between environments. They ensure that the application and its dependencies are packaged together and isolated from the underlying infrastructure. This allows the application to run quickly and reliably from one computing environment to another.

We will start by installing Docker, a platform for building and running containers, on your local system. We will run simple Docker images and learn the basic Docker commands. We will then build our first Docker image containing a simple Python application. We will learn how to define a Dockerfile to efficiently specify the environment and dependencies for our application. We will then run our image as a container and explore how to access the application and check its logs.

Containers are a key technology for modern software deployment. They are lightweight, portable, and scalable, allowing you to build and ship applications faster. The concepts and skills you will learn in this chapter will provide a strong foundation for working with containers and deploying data applications. By the end of this chapter, you will be ready to start building and deploying your own containerized data processing jobs, APIs, and data engineering tools.

In this chapter, we're going to cover the following main topics:

- Container architecture
- Installing Docker
- Getting started with Docker images
- Building your own image

Technical requirements

For this chapter, you should have Docker installed. Also, a computer with a minimum of 4 GB of RAM (8 GB is recommended) is required, as Docker can really consume a computer's memory.

The code for this chapter is available on GitHub. Please refer to `https://github.com/PacktPublishing/Bigdata-on-Kubernetes` and access the `Chapter01` folder.

Container architecture

Containers are an operating system-level virtualization method that we can use to run multiple isolated processes on a single host machine. Containers allow applications to run in an isolated environment with their own dependencies, libraries, and configuration files without the overhead of a full **virtual machine** (**VM**), which makes them lighter and more efficient.

If we compare containers to traditional VMs, they differ in a few ways. VMs virtualize at the hardware level, creating a full virtual operating system. Containers, on the other hand, virtualize at the operating system level. Because of that, containers share the host system's kernel, whereas VMs each have their own kernel. This allows containers to have much faster startup times, typically in milliseconds compared to minutes for VMs (it is worth noting that in a Linux environment, Docker can leverage the capabilities of a Linux kernel directly. While running in a Windows system, however, it runs in a lightweight Linux VM that is still lighter than a full VM).

Also, containers have better resource isolation as they only isolate the application layer, whereas VMs isolate an entire operating system. Containers are immutable infrastructure, making them more portable and consistent as updates create new container images (versions) rather than updating in place.

Due to these differences, containers allow higher density, faster startup times, and lower resource usage compared to VMs. A single server can run dozens or hundreds of containerized applications isolated from each other.

Docker is one of the most popular container platforms that provides tools to build, run, deploy, and manage containers. Docker architecture consists of the Docker client, Docker daemon, Docker registry, and Docker images.

The Docker client is a **command-line interface** (**CLI**) client used to interact with the Docker daemon to build, run, and manage containers. This interaction occurs through a REST API.

The Docker daemon is a background service that runs on the host machine and manages building, running, and distributing containers. It is the base for all the containers to run on.

The Docker registry is a repository to host, distribute, and download Docker images. Docker Hub is the default public registry with many pre-built images, but cloud providers usually have their own private container registry as well.

Finally, Docker images are read-only templates used to create Docker containers. Images define the container environment, dependencies, operating system, environment variables, and everything that a container needs to run.

Figure 1.1 shows the difference between an application running in a VM and an application running in a container.

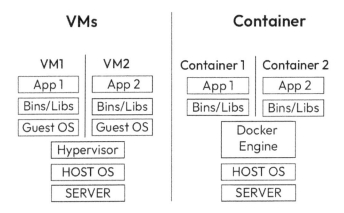

Figure 1.1 – VMs versus containers

Figure 1.2 shows how a container runs with separate layer levels. There is the shared kernel at the bottom. On top of that, we have as many operating systems as we need. On top of the Debian OS layer, we see a Java 8 image and an NGINX image. The Java 8 layer is shared by three containers, one of them with only the image information and two using another image, Wildfly. The figure demonstrates why the container architecture is so efficient in sharing resources and lightweight because it is built upon layers of libraries, dependencies, and applications that will run isolated from each other.

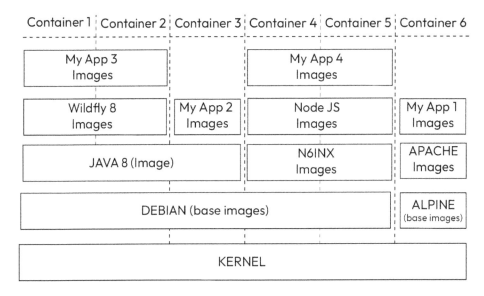

Figure 1.2 – Container layers

Now, let's get to it. In the next section, you will learn how to install Docker and run your first Docker CLI commands.

Installing Docker

To get started with Docker, you can install it by using the package manager for your Linux distribution or install Docker Desktop for Mac/Windows machines.

Windows

To use Docker Desktop on Windows, you must turn on the WSL 2 feature. Refer to this link for detailed instructions: `https://docs.microsoft.com/en-us/windows/wsl/install-win10`.

After that, you can install Docker Desktop as follows:

1. Go to `https://www.docker.com/products/docker-desktop` and download the installer.

2. When the download is ready, double-click the installer and follow the prompts.

    You should ensure that the **Use WSL 2 instead of Hyper-V** option is selected on the **Configuration** page. This is the recommended usage. (If your system does not support WSL 2, this option will not be available. You can still run Docker with Hyper-V, though.)

3. After the installation is finished, close to complete and start Docker Desktop.

If you have any doubts, refer to the official documentation: `https://docs.docker.com/desktop/install/windows-install/`.

macOS

The installation of Docker Desktop on macOS is quite simple:

1. Go to `https://www.docker.com/products/docker-desktop` and download the installer for macOS.

2. Double-click the installer and follow the prompts to install Docker Desktop.

3. Once the installation completes, Docker Desktop will start automatically.

Docker Desktop runs natively on macOS using the HyperKit VM and does not need additional configuration. When Docker Desktop starts for the first time, it will prompt you to authorize it for drive access. Authorize Docker Desktop to allow it to access files on your filesystem.

Linux

Installing Docker on Linux-based systems is very straightforward. You can use your Linux distribution package manager to do that with just a few commands. For Ubuntu, for instance, the first thing is to remove any older versions of Docker that you previously had on the machine:

```
$ sudo apt-get remove docker docker-engine docker.io containerd runc
```

You can install Docker from the default `apt` repository using the following:

```
$ sudo apt install docker.io
```

This will install a slightly older version of Docker. If you want the latest version, check the official Docker website (`https://docs.docker.com/desktop/install/linux-install/`) and follow the instructions.

If you want to use Docker without having to `sudo` commands, run the following commands:

```
$ sudo groupadd docker
$ sudo usermod -aG docker <YOUR_USERNAME>
```

Now, let's get hands-on practice with Docker.

Getting started with Docker images

The very first Docker image we can run is the `hello-world` image. It is often used to test whether Docker is correctly installed and running.

hello-world

After the installation, open the terminal (Command Prompt in Windows) and run the following:

```
$ docker run hello-world
```

This command will pull the `hello-world` image from the Docker Hub public repository and run the application in it. If you can run it successfully, you will see this output:

```
Unable to find image 'hello-world:latest' locally
latest: Pulling from library/hello-world
70f5ac315c5a: Pull complete
Digest: sha256:88ec0acaa3ec199d3b7eaf73588f4518c25
f9d34f58ce9a0df68429c5af48e8d
Status: Downloaded newer image for hello-world:latest


Hello from Docker!
This message shows that your installation appears to be working
correctly.
```

```
To generate this message, Docker took the following steps:
 1. The Docker client contacted the Docker daemon.
 2. The Docker daemon pulled the "hello-world" image from the Docker
Hub.
 3. The Docker daemon created a new container from that image which
runs the executable that produces the output you are currently
reading.
 4. The Docker daemon streamed that output to the Docker client, which
sent it to your terminal.

To try something more ambitious, you can run an Ubuntu container with:
 $ docker run -it ubuntu bash

Share images, automate workflows, and more with a free Docker ID:
 https://hub.docker.com/

For more examples and ideas, visit:
 https://docs.docker.com/get-started/
```

Congratulations! You just ran your first Docker image! Now, let's try something a little more ambitious.

NGINX

NGINX is a well-known open source software for web serving, reverse proxying, and caching. It is widely used in Kubernetes-based architectures.

Different from the hello-world application (which behaves like a job execution), NGINX behaves like a service. It opens a port and keeps listening for user requests. We can start by searching for the available NGINX images in Docker Hub using the following:

```
$ docker search nginx
```

The output will show several available images. Usually, the first in the list is the official image. Now, to set up a running NGINX container, we can use the following command:

```
$ docker pull nginx:latest
```

This will download the current latest version of the image. The latest keyword after the colon stands for the "tag" of this image. To install a specific version (recommended), specify the tag like this:

```
$ docker pull nginx:1.25.2-alpine-slim
```

You can visit https://hub.docker.com/_/nginx to check for all the available tags.

Now, to run the container, you should specify which version of the image you want to use. The following command will do the trick:

```
$ docker run --name nginxcontainer -p 80:80 nginx:1.25.2-alpine-slim
```

You will start to see NGINX logs in the terminal. Then, open your preferred browser and type http://localhost/. You should see this message (*Figure 1.3*):

Welcome to nginx!

If you see this page, the nginx web server is successfully installed and working. Further configuration is required.

For online documentation and support please refer to nginx.org. Commercial support is available at nginx.com.

Thank you for using nginx.

Figure 1.3 – The nginx default output in the browser

The docker run command has a few important parameters. --name defines the name of the container that will run. If you don't define a name, Docker will automatically choose a name for it (trust me, it can be very creative). -p connects a port on your machine (port 80) to a port inside the container (also 80). If you do not open this port, you won't be able to reach the container's running application.

After your test is successful, get back to the terminal running the container and press "CTRL + C" to stop the container. After it stops, it will still be there, although not running. To remove the container, use the following:

```
$ docker rm nginxcontainer
```

If you are in doubt, you can see all the running and stopped containers with this command:

```
$ docker ps -a
```

We also can see all the locally available images with this command:

```
$ docker images
```

In my case, the output shows three images: hello-world and two NGINX images, one of them with the latest tag and the other with the 1.25.2-alpine-slim tag. All images and their respective versions will show.

Julia

In this last example, we will learn how to use technology that is not installed in our machine by interacting with running containers. We will run a container with a new and efficient programming language for data science called Julia. To do that, execute the following command:

```
$ docker run -it --rm julia:1.9.3-bullseye
```

Note that the docker run command looks for a local image. If it's not downloaded, Docker will automatically pull the image from Docker Hub. With the preceding command, we will start an interactive session in a Julia 1.9.3 container. The -it parameters allow us to use it interactively. The --rm parameter states that the container will automatically be removed after it is stopped, so we don't have to manually remove it.

After the container is up and running, let's play with a simple custom function to calculate two descriptive statistics: a mean and a standard deviation. You will see a Julia's logo in the terminal, and you can use the following:

```
$ using Statistics
$ function descriptive_statistics(x)
    m = mean(x)
    sd = std(x)
    return Dict("mean" => m, "std_dev" => sd)
  end
```

After defining the function, we will run it with a small array of random numbers:

```
$ myvector = rand(5)
$ descriptive_statistics(myvector)
```

You should see the proper output on the screen. Congratulations! You have just used the Julia programming language without having to install or configure it on your computer with Docker! To exit the container, use the following:

```
$ exit()
```

As we used the --rm parameter, if we run the docker ps -a command, we will see that it has been automatically removed.

Building your own image

Now, we will customize our own container images for running a simple data processing job and an API service.

Batch processing job

Here is a simple Python code for a batch processing job:

run.py

```
import pandas as pd

url = 'https://raw.githubusercontent.com/jbrownlee/Datasets/master/
pima-indians-diabetes.data.csv'

df = pd.read_csv(url, header=None)
df["newcolumn"] = df[5].apply(lambda x: x*2)

print(df.columns)
print(df.head())
print(df.shape)
```

This Python code loads a CSV dataset from a URL into a pandas DataFrame, adding a new column by multiplying an existing column by 2 and then printing out some information about the DataFrame (column names, first five rows, and size of the DataFrame). Type this code using your favorite code editor and save the file with the name run.py.

Normally, we test our code locally (whenever possible) to be sure it is working. To do that, first, you need to install the pandas library:

```
pip3 install pandas
```

Then, run the code with the following:

```
python3 run.py
```

If all goes well, you should see an output like this:

```
Index([0, 1, 2, 3, 4, 5, 6, 7, 8, 'newcolumn'], dtype='object')
    0    1   2   3    4     5      6   7  8  newcolumn
0   6  148  72  35    0  33.6  0.627  50  1       67.2
1   1   85  66  29    0  26.6  0.351  31  0       53.2
2   8  183  64   0    0  23.3  0.672  32  1       46.6
3   1   89  66  23   94  28.1  0.167  21  0       56.2
4   0  137  40  35  168  43.1  2.288  33  1       86.2
(768, 10)
```

Now, we are ready to package our simple processing job into a container. Let's start by defining a Dockerfile:

Dockerfile_job

```
FROM python:3.11.6-slim

RUN pip3 install pandas

COPY run.py /run.py

CMD python3 /run.py
```

Those are the only four lines we need to define a working container. The first line specifies the base image to use, which is a slim version of Python 3.11.6. This is a Debian-based OS that already has Python 3.11.6 installed, which can save us a lot of time. Using a slim image is very important to keep the container size small and optimize transfer time and storage costs (when it's the case).

The second line installs the pandas library. The third line copies the local run.py file into the container. Finally, the last line sets the default command to run when the container starts to execute the /run.py script. After you are done defining the code, save it as Dockerfile_job (no .extension). Now, it's time to build our Docker image:

```
docker build -f Dockerfile_job -t data_processing_job:1.0 .
```

The docker build command builds an image according to the Dockerfile instructions. Usually, this command expects a file named Dockerfile. Since we are working with a filename different from expected, we must tell Docker which file to use with the -f flag. The -t flag defines a tag for the image. It is composed of a name and a version, separated by a colon (:). In this case, the name we set for the image is data_processing_job and a 1.0 version. The last parameter to this command is the path where code files are located. Here, we set the current folder with a dot (.). This dot is very easy to forget, so be careful!

After the build is finished, we can check the locally available images with this command:

```
docker images
```

You should see the first line of the output showing your recently built data processing image:

```
REPOSITORY      TAG  IMAGE ID      CREATED        SIZE
data_process... 1.0  39bae1eb068c  6 minutes ago  351MB
```

Now, to run our data processing job from inside the container, use this command:

```
docker run --name data_processing data_processing_job:1.0
```

The docker run command runs the specified image. The --name flag defines the name of the container as data_processing. After you start running the container, you should see the same output as before:

```
Index([0, 1, 2, 3, 4, 5, 6, 7, 8, 'newcolumn'], dtype='object')
    0   1   2   3    4     5      6   7  8  newcolumn
0   6  148  72  35    0  33.6  0.627  50  1      67.2
1   1   85  66  29    0  26.6  0.351  31  0      53.2
2   8  183  64   0    0  23.3  0.672  32  1      46.6
3   1   89  66  23   94  28.1  0.167  21  0      56.2
4   0  137  40  35  168  43.1  2.288  33  1      86.2
(768, 10)
```

Finally, don't forget to remove the exited containers from your environment:

```
docker ps -a
docker rm data_processing
```

Congrats! You have run your first job using containers. Now, let's move to another type of containerized application: a service.

API service

In this section, we will use **FastAPI** to develop a simple API with Python. Open a Python script in your favorite code editor, create a folder named app, and create a Python script named main.py.

In the script, first, we import FastAPI and the random module:

```
from fastapi import FastAPI
import random
```

Next, we create an instance of the FastAPI app:

```
app = FastAPI()
```

The next code block defines a route using the @app.get decorator:

```
@app.get("/api")
async def root():
    return {"message": "You are doing great with FastAPI..."}
```

The @app.get decorator indicates that this is a GET endpoint. This function is defined to answer at the "/api" route. It just returns a pleasant message on request of the route.

The next code chunk defines a route, `"/api/{name}"`, where `name` is a parameter that will be received in the request. It returns a greeting message with the given name:

```
@app.get("/api/{name}")
async def return_name(name):
    return {
        "name": name,
        "message": f"Hello, {name}!"
    }
```

The last code block defines a `"/joke"` route. This function returns a (very funny!) random joke from the list of jokes previously defined. Feel free to replace them with your own cool jokes:

```
@app.get("/joke")
async def return_joke():
    jokes = [
        "What do you call a fish wearing a bowtie? Sofishticated.",
        "What did the ocean say to the beach? Nothing. It just waved",
        "Have you heard about the chocolate record player? It sounds
pretty sweet."

    ]

    return {
        "joke": random.choice(jokes)
    }
```

It is important to notice that every function returns a response in JSON format. This is a very common pattern with APIs. For the whole Python code, refer to the book's GitHub repository (https://github.com/PacktPublishing/Bigdata-on-Kubernetes).

Before we build the Docker image, it is advised to test the code locally (whenever possible). To do this, you must install the `fastapi` and `uvicorn` packages. Run this command in the terminal:

```
pip3 install fastapi uvicorn
```

To run the API, use the following:

```
uvicorn app.main:app --host 0.0.0.0 --port 8087
```

If all goes well, you will see an output like this in the terminal:

```
INFO:    Started server process [14417]
INFO:    Waiting for application startup.
INFO:    Application startup complete.
INFO:    Uvicorn running on http://0.0.0.0:8087 (Press CTRL+C to quit)
```

This command runs the API service locally on port 8087. To test it, open a browser and access http://localhost:8087/api. You should see the programmed message on the screen. Test the http://localhost:8087/api/<YOUR_NAME> and http://localhost:8087/joke endpoints as well.

Now that we know everything is working fine, let's package the API in a Docker image. To do that, we will build a simple Dockerfile. To optimize it, we will use the alpine linux distribution, an extremely lightweight base OS. In the root folder of your project, create a new file named Dockerfile (no .extension). This is the code we will use for this image:

Dockerfile

```
FROM python:3.11-alpine

RUN pip3 --no-cache-dir install fastapi uvicorn

EXPOSE 8087

COPY ./app /app

CMD uvicorn app.main:app --host 0.0.0.0 --port 8087
```

The first line imports a Python container based on the alpine Linux distribution. The second line installs fastapi and uvicorn. The third line informs Docker that the container will listen on port 8087 at runtime. Without this command, we would not be able to access the API service. The fourth line copies all the code inside our local /app folder to a /app folder inside the container. Finally, the CMD command specifies the command to run when the container starts. Here, we are starting the uvicorn server to run our FastAPI application. After uvicorn, we state a location pattern of folder.script_name:FastAPI_object_name to tell FastAPI where to look for the API process object.

When this Dockerfile is built into an image, we will have a containerized Python application configured to run a FastAPI web server on port 8087. The Dockerfile allows us to package up the application and its dependencies into a standardized unit for deployment. To build the image, run the following:

```
docker build -t my_api:1.0 .
```

No need to specify the -f flag here since we are using a Dockerfile with the default name. And remember the dot at the end of the line!

Now, we run the container with a slightly different set of parameters:

```
docker run -p 8087:8087 -d --rm --name api my_api:1.0
```

The -p parameter sets that we will open port 8087 in the server (in this case, your computer) to port 8087 in the container. If we don't set this parameter, there is no way to communicate with the container whatsoever. The -d parameter runs the container in *detached* mode. The terminal will not be showing container logs but it will be available for use while the container is running in the background. The --rm parameter sets the container to be automatically removed after it is finished (very handy). Finally, --name sets the name for the container as api.

We can check whether the container is correctly running with the following:

```
docker ps -a
```

If you need to check the logs to a container, use the following:

```
docker logs api
```

You should see an output similar to this:

```
INFO:    Started server process [1]
INFO:    Waiting for application startup.
INFO:    Application startup complete.
INFO:    Uvicorn running on http://0.0.0.0:8087 (Press CTRL+C to quit)
```

Now, we can test our API endpoints in the browser with the same links shown before (http://localhost:8087/api, http://localhost:8087/api/<YOUR_NAME> and http://localhost:8087/joke).

Congrats! You are running your API service from inside a container. This is a completely portable and self-contained application that can be deployed anywhere.

To stop the API service, use the following:

```
docker stop api
```

To check that the stopped container has been automatically removed, use the following:

```
docker ps -a
```

Nicely done!

Summary

In this chapter, we covered the fundamentals of containers and how to build and run them using Docker. Containers provide a lightweight and portable way to package applications and their dependencies so they can run reliably across environments.

You learned about key concepts such as images, containers, Dockerfiles, and registries. We installed Docker and ran simple containers such as NGINX and Julia to get hands-on experience. You built your own containers for a batch processing job and API service, defining Dockerfiles to package dependencies.

These skills allow you to develop applications and containerize them for smooth deployment anywhere. Containers are super useful as they ensure your software runs exactly as intended every time.

In the next chapter, we will look at orchestrating containers using Kubernetes to easily scale, monitor, and manage containerized applications. We will take a look at the most important Kubernetes concepts and components and learn how to implement them with YAML files (manifests).

2
Kubernetes Architecture

Understanding Kubernetes architecture is crucial to properly leverage its capabilities. In this chapter, we will go over the main components and concepts that make up a Kubernetes cluster. Getting familiar with these building blocks will allow you to understand how Kubernetes works under the hood.

We will start by looking at the different components that make up a Kubernetes cluster – the control plane and the worker nodes. The control plane—made up of components such as the API server, controller manager, and etcd—is responsible for managing and maintaining the desired state of the cluster. The worker nodes run your containerized applications in pods.

After covering the cluster architecture, we will dive into the main Kubernetes abstractions and API resources such as pods, deployments, StatefulSets, services, ingress, and persistent volumes. These resources allow you to declare the desired state of your applications and have Kubernetes reconcile the actual state to match it. Understanding these concepts is key to being able to deploy and manage applications on Kubernetes.

We will also look at supporting resources such as ConfigMaps and Secrets, which allow you to separate configuration from code. Jobs provide support for batch workloads.

By the end of this chapter, you will have a solid understanding of how a Kubernetes cluster is put together and how you can leverage its capabilities by utilizing its API resources. This will enable you to start deploying your own applications and managing them efficiently.

We'll be covering these concepts under the following main topics:

- Cluster architecture
- Pods
- Deployments
- StatefulSets
- Jobs
- Services

- Ingress and ingress controllers

- Gateway

- Persistent volumes

- ConfigMaps and Secrets

Technical requirements

There are no technical requirements for this chapter. All code presented here is generic and practical executable examples will be given in *Chapter 3*.

Kubernetes architecture

Kubernetes is a cluster architecture. This means that in a full production environment, you will usually have several machines running your workloads simultaneously to create a reliable and scalable architecture. (Note that Kubernetes can run on one machine also, which is great for testing but misses the whole point for production.)

To coordinate cluster functionalities, Kubernetes has two main feature groups: the **control plane** responsible for cluster management and the **node components** that communicate with the control plane and execute tasks in the worker machines. *Figure 2.1* shows a representation of the whole system.

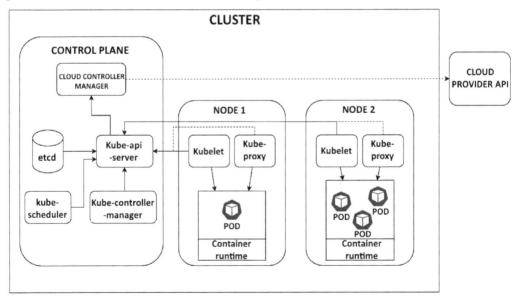

Figure 2.1 – Kubernetes architecture

Let's take a closer look at each group and its components.

Control plane

The main components of the control plane are the following:

- kube-apiserver
- etcd
- kube-scheduler
- kube-controller-manager

When running a cloud-based Kubernetes cluster, we also have another component called cloud-controller-manager.

kube-apiserver

The Kubernetes API is exposed to the administrator with **kube-apiserver**. It can be considered the "frontend" of the control plane. It is through the API server that we will interact with Kubernetes, sending instructions to the cluster or getting data from it. It is highly scalable and scales horizontally (deploying more worker nodes to the cluster).

etcd

Kubernetes utilizes etcd, a distributed key-value database, to persistently store all cluster data and state. Etcd serves as the backing store for the Kubernetes API server, providing a secure and resilient foundation for the orchestration of containers across nodes in a cluster. By leveraging etcd's capabilities for consistency, high availability, and distribution, Kubernetes can reliably manage the desired state of applications and infrastructure. etcd is fault tolerant even if the failure happens in a leader node of the cluster.

kube-scheduler

kube-scheduler is responsible for distributing work or containers across multiple nodes. It watches for newly created pods that are not assigned to any node and selects a node for them to run on. To make scheduling decisions, kube-scheduler analyzes individual and collective available resources, hardware/software/policy constraints, affinity and anti-affinity instructions, deadlines, data locality and eventual interferences between workloads.

kube-controller-manager

kube-controller-manager runs controllers that regulate behavior in the cluster, such as node controllers, job controllers, EndpointSlice controllers, and ServiceAccount controllers. The controllers reconcile the desired state with the current state.

cloud-controller-manager

cloud-controller-manager interacts with the underlying cloud providers and sets up cloud-based networking services (such as networking routes and load balancers). It separates components that interact with the cloud provider from the components that run only inside the cluster. This controller manager only runs controllers that are specific to the cloud provider in use, thus, if you are running a test local Kubernetes instance, cloud-controller-manager will not be used, since it only deals with cloud-based services.

Next, let's have a look at the node components.

Node components

Node components are present in every single worker node of the cluster and are responsible for communicating with the control plane, running and maintaining workloads, and providing a runtime environment. The main components are as follows:

- Container runtime
- kubelet
- kube-proxy

Container runtime

A container runtime is the underlying software that is responsible for running containers. Kubernetes supports several container runtimes, but the most common ones are Docker and containerd. The container runtime is responsible for pulling the images from the registries, running the containers, and managing containers' lifecycles.

kubelet

kubelet is the primary node agent that watches the assigned pods and ensures containers are running and healthy. It interacts with the container runtime to pull images and run containers.

kube-proxy

kube-proxy is a network proxy and load balancer that implements Kubernetes networking services on each node by maintaining network rules and performing connection forwarding.

Now, let's move our attention to the Kubernetes components we will use to build our workloads.

Pods

Pods are the smallest deployable units in Kubernetes and represent a single instance of an application. Pods contain one or more containers (although the most common case is to have just one container inside a pod). When multiple containers live inside a pod, they are guaranteed to be co-located on the same node and can share resources.

Pods provide two main benefits:

- **Resource sharing and communication**: Containers within the same pod can communicate over `localhost` and share resources such as volumes. This facilitates easy communication between related containers. It is important to notice, though, that this is an advanced use case and should be used only when your containers are tightly coupled. We regularly use pods for single-container deployments.

- **Management and deployment**: Pods are the units that get deployed, scaled, and managed in Kubernetes. You don't directly create or manage the containers within pods. This entire process is fully automated.

Usually, you do not define pods directly. Pods can be created and managed though other resources such as deployments, jobs, and `StatefulSets`. Nevertheless, you can define a pod with a proper YAML file manifest.

Manifests are the basic way of telling Kubernetes about the desired state of any object or giving it instructions about how to act or deploy anything. It is written as a `.yaml` file. **YAML files** are often used for configuration. They are very close to JSON files, but they are more readable since they rely on indentation for code hierarchy structure rather than brackets and braces. The following is an example of a manifest to deploy a single pod:

pod.yaml

```
apiVersion: v1
kind: Pod
metadata:
    name: myapp-pod
    labels:
    app: myapp
spec:
    containers:
    - name: myapp-container
    image: myimage:latest
    ports:
    - containerPort: 80
```

Let's have a closer look at each part of this manifest:

- `apiVersion`: The Kubernetes API version for the objects in this manifest. For pods this is `v1`.

- `kind`: The type of object being created, which is `Pod` for this manifest.

- **metadata**: This section contains metadata for the pod, such as the name and labels. The name is a unique identifier. Labels will be particularly important in the future as they serve as identifiers for other Kubernetes resources such as deployments and services.

- **spec**: This section defines the desired state of the pod including its containers.

- **containers**: Specifies the container(s) that run inside the pod. Includes the image, ports, etc.

- **image**: The Docker image to use for the container. It can be a public or private image.

- **ports**: Defines the ports exposed by the container.

This covers the basic structure of a pod manifest. Pods provide a simple way to deploy and manage containers in Kubernetes. Now that we have discussed pods and how to define them, let's discuss a way of automating more complex pod structures with deployments.

Deployments

Deployments are one of the most important resources in Kubernetes for running applications. They provide a declarative way to manage pods and replicas.

A deployment defines the desired state for your application, including the container image, number of replicas, resource limits, and more. The Kubernetes control plane works to match the actual state of your cluster to the desired state in the deployment.

For example, here is a simple deployment manifest:

deployment.yaml

```
apiVersion: apps/v1
kind: Deployment
metadata:
    name: myapp-deployment
spec:
    replicas: 3
    selector:
    matchLabels:
      app: myapp
    template:
    metadata:
      labels:
        app: myapp
    spec:
      containers:
      - name: myapp
```

```
        image: nginx:1.16
        ports:
        - containerPort: 80
```

Let's break this down section by section:

- `apiVersion`: This specifies the Kubernetes API version for the Deployment resource. We want the `apps/v1` version which includes Deployments.

- `kind: Deployment`: The type of resource we are creating, in this case, a Deployment.

- `metadata`: Standard metadata for the resource like a unique name.

- `spec`: The specification for the Deployment. This defines the desired state.

- `replicas: 3`: We want three pod replicas to be running. Kubernetes will maintain this number of pods.

- `selector`: Used to match pods managed by this Deployment. Pods will be selected based on the label selector.

- `template`: The template for the pods that will be created. It defines the pod specifications. Note that the deployment will relate to the label specified.

- `spec: containers`: Pod spec including the container(s) to run.

- `image: nginx:1.16`: The container image to use.

- `ports`: Ports exposed by the container.

When this Deployment is applied, Kubernetes will launch three pods matching the template, each running an Nginx container. The Deployment controller will monitor the pods and ensure the desired state matches the actual state, restarting pods if needed.

Deployments provide powerful capabilities for running scalable and resilient applications on Kubernetes. Using declarative configuration makes deployments easy. Next, we will discuss a different approach for managing pods and replicas: StatefulSets.

StatefulSets

StatefulSets are a Kubernetes resource used to manage stateful applications such as databases. They are similar to Deployments but are designed to handle stateful workloads that require persistent storage and unique network identifiers.

A StatefulSet manages Pods that contain stateful applications (applications that must keep track of data for other applications or other user sessions). The Pods in a StatefulSet have a sticky, unique identity that persists across rescheduling. This allows each Pod to maintain its state when restarted or rescheduled onto a new node. This makes StatefulSets ideal for stateful apps such as databases that require data persistence. Deployments, on the other hand, are designed for stateless workloads and provide identical Pods with no persistent storage. Thus, they are better for stateless web apps.

StatefulSets operate by creating PersistentVolumes (which will be covered later in this chapter) for each Pod to mount. This ensures data persists across Pod restarts. StatefulSets also provide a unique hostname and stable network ID per Pod using a predictable naming convention.

Here is an example of a statefulset manifest for deploying a MySQL database:

statefulset.yaml

```yaml
apiVersion: apps/v1
kind: StatefulSet
metadata:
    name: mysql
spec:
    serviceName: "mysql"
    replicas: 1
    selector:
    matchLabels:
      app: mysql
    template:
    metadata:
      labels:
        app: mysql
    spec:
      containers:
      - name: mysql
        image: mysql:5.7
        ports:
        - containerPort: 3306
          name: mysql
        volumeMounts:
        - name: mysql-persistent-storage
          mountPath: /var/lib/mysql
    volumeClaimTemplates:
    - metadata:
      name: mysql-persistent-storage
    spec:
      accessModes: [ "ReadWriteOnce" ]
      storageClassName: "my-storage-class"
      resources:
        requests:
          storage: 1Gi
```

This manifest creates a StatefulSet for MySQL with one replica. It uses a `volumeClaimTemplate` to dynamically provision a PersistentVolume for each Pod. The MySQL data will be persisted in the `/var/lib/mysql` path.

Each Pod gets a unique name, such as `mysql-0`, and a stable hostname. If the Pod gets rescheduled, it will remount its PersistentVolume to continue running statefully.

In this way, StatefulSets provide powerful stateful management for databases and other stateful apps in Kubernetes. They ensure persistence, stable networking, ordered deployment, and graceful scaling. Next, we will continue with a discussion about Kubernetes jobs.

Jobs

Jobs are a fundamental resource type in Kubernetes used to run batch processes that run to completion. Unlike long-running services such as web servers, jobs are intended to terminate when the batch process finishes.

A job creates one or more pods that run a defined workload and then terminates when the workload is complete. This is useful for tasks such as data processing, machine learning training, or any finite computation.

To create a job, you define a Job resource in a YAML manifest like this:

job.yaml

```
apiVersion: batch/v1
kind: Job
metadata:
    name: myjob
spec:
    template:
    spec:
      containers:
      - name: myjob
        image: busybox
        command: ['sh', '-c', 'echo Hello Kubernetes! && sleep 30']
      restartPolicy: Never
    backoffLimit: 4
```

Let's have a closer look at this code, part by part:

- `apiVersion` ' `batch/v1` for jobs

- `kind` ' `Job`

- `metadata.name` ' Name of the job

- `spec.template` ' Pod template defining the container(s) to run the same way we saw in the previous resource definitions

- `spec.template.spec.restartPolicy` ' Set to `Never` since jobs shouldn't restart

- `spec.backoffLimit` ' Optional limit on failed job retries

The pod template under `spec.template` defines the container(s) to run just like a pod manifest. You can specify the image, commands, environment variables, and so on. An important setting is the `restartPolicy`, which should be `Never` for jobs. This ensures pods are not restarted if they fail or exit. The `backoffLimit` is optional and specifies the number of times a failed job pod can be retried. The default is **6**. Set this lower if jobs should not retry too many times on failure.

When you create the job, Kubernetes will schedule one or more pods matching the template to run your workload. As the pods finish, Kubernetes will track their status and know when the job is completed. You can view job status and pod logs to monitor progress. Jobs make it easy to run batch computational workloads on Kubernetes. In the next section, we will take a look at Kubernetes services.

Services

Services provide stable endpoints to access pods running in a cluster, thus exposing our applications to users online. They allow pods to die and replicate without interrupting access to the applications running in those pods. There are several types of services in Kubernetes. We will discuss three of them in detail: ClusterIP, NodePort, and load balancer.

ClusterIP Service

A ClusterIP service provides an IP address that is only accessible inside the cluster. This IP does not change for the lifetime of the service, providing a stable endpoint to access the pods. Here is an example ClusterIP service manifest:

service_clusterip.yaml

```
apiVersion: v1
kind: Service
metadata:
    name: my-service
spec:
    type: ClusterIP
    selector:
    app: my-app
    ports:
    - protocol: TCP
     port: 80
     targetPort: 9376
```

This manifest creates a service called `my-service` that will forward requests to pods with the label `app: my-app` on `port 80`. Requests will be forwarded to `port 9376` on the target pods. The ClusterIP will not change while this service exists.

NodePort Service

A NodePort service makes an internal ClusterIP service accessible externally through a port allocated on each node. The NodePort is allocated from a configured range (by default, `30000-32767`) and will be the same on every node. Traffic to `<NodeIP>:<NodePort>` will be forwarded to the ClusterIP service. Here is an example:

service_nodeport.yaml

```
apiVersion: v1
kind: Service
metadata:
    name: my-service
spec:
    type: NodePort
    selector:
    app: my-app
    ports:
    - port: 80
      targetPort: 9376
      nodePort: 30007
```

This exposes the internal ClusterIP on port `30007` on every node. Requests to `<NodeIP>:30007` will be forwarded to the service.

LoadBalancer Service

A load balancer service provisions an external load balancer to expose the service to external traffic. A ClusterIP exposes the service on an internal IP address within the Kubernetes cluster. This makes the service only reachable within the cluster. Load balancer, on the other hand, exposes the service externally using the cloud provider's load balancer implementation. This makes the service reachable from outside the Kubernetes cluster.

The load balancer implementation depends on the environment. For example, on AWS, this would create an **Elastic Load Balancer** (**ELB**), an AWS service to provide a managed load balancer. Here is an example:

service_loadbalancer.yaml

```
apiVersion: v1
kind: Service
metadata:
    name: my-service
spec:
```

```
    selector:
    app: my-app
    ports:
    - protocol: TCP
      port: 80
      targetPort: 9376
    type: LoadBalancer
```

This creates a load balancer and assigns an external IP address. Traffic to the external IP is forwarded to the internal ClusterIP service. Next, we will discuss a different way of defining services with Ingress and Ingress Controller.

Ingress and Ingress Controller

An **Ingress** resource defines rules for external connectivity to Kubernetes services. It enables inbound HTTP and HTTPS connections to reach services running within the cluster. Traffic routing is controlled by rules defined on the Ingress resource. For an ingress to be able to run, you need to have a running ingress controller on Kubernetes.

An **Ingress controller** is responsible for fulfilling the Ingress, usually with a load balancer. It watches for Ingress resources and configures the load balancer accordingly. Different load balancers require different Ingress controller implementations.

Some examples of Ingress controllers include the following:

- NGINX Ingress Controller: Uses NGINX as a load balancer and reverse proxy. It is one of the most common and fully featured controllers.

- HAProxy Ingress Controller: Uses HAProxy for load balancing. Provides high performance and reliability.

- Traefik Ingress Controller: A cloud-native controller that integrates with Let's Encrypt for automatic HTTPS certificate generation.

- AWS ALB Ingress Controller: Uses the AWS **Application Load Balancer** (**ALB**). Integrates natively with other AWS services.

The Ingress resource contains two main parts – a backend and rules. The backend specifies the default service to route unmatched requests. The rules contain a set of paths and the services to route them to:

ingress.yaml

```
apiVersion: networking.k8s.io/v1
kind: Ingress
metadata:
```

```
            name: example-ingress
    spec:
        backend:
        service:
          name: example-service
          port:
            number: 80
        rules:
        - http:
            paths:
              - path: /
                pathType: Prefix
                backend:
                  service:
                    name: example-service
                    port:
                      number: 80
```

In this example, requests to the root path / will be routed to the `example-service` on port `80`. The `pathType:` prefix indicates that any subpath should also be routed to the service.

Multiple rules can be defined to route different paths to different services:

```
    spec:
        rules:
        - http:
            paths:
              - path: /foo
                backend:
                  service:
                    name: foo-service
                    port:
                      number: 80
        - http:
            paths:
              - path: /bar
                backend:
                  service:
                    name: bar-service
                    port:
                      number: 80
```

With the preceding code, requests to `/foo` will go to `foo-service` and requests to `/bar` will go to `bar-service`.

In some cases, we have to configure secure connections with in-transit encryption. When this is the case, we can configure advanced encryption options in Ingress controllers using annotations:

```
metadata:
    annotations:
        nginx.ingress.kubernetes.io/ssl-redirect: "false"
```

Host-based routing can also be configured by specifying host names:

```
spec:
    rules:
    - host: foo.example.com
    http:
      paths:
      - path: /
        backend:
            serviceName: foo-service
            servicePort: 80
    - host: bar.example.com
    http:
      paths:
      - path: /
        backend:
            serviceName: bar-service
            servicePort: 80
```

Now `foo.example.com` will route to `foo-service` and `bar.example.com` to `bar-service`.

In summary, Ingress provides a way to intelligently route HTTP and HTTPS traffic to services in a Kubernetes cluster. Ingress controllers handle the actual load balancing and reverse proxy functionality. Common use cases for Ingress include exposing services to external users and handling TLS/SSL. Careful Ingress configuration is crucial for production-grade Kubernetes deployments.

It is important to note that the ingress API is frozen. That means that this API will not be receiving any more updates. It is replaced by the Gateway API. Nevertheless, it is important to know it since a lot of the big data tools that we will use in this book are still deployed using Ingress instructions. Now, let's move to the Gateway API and understand how it works.

Gateway

The Gateway API is a Kubernetes API that provides a way to dynamically configure load balancing and service mesh capabilities on Kubernetes. The Gateway API allows defining routes and policies to manage external traffic to Kubernetes services in a centralized, declarative way.

The main resources in Gateway API are the following:

- `GatewayClass` ' Defines a set of gateways with a common configuration and behavior. It is like the concept of StorageClass for persistent volumes.

- `Gateway` ' Defines a set of routes for a given hostname. This binds GatewayClass, TLS certificate, and other configurations to a set of routes.

- `HTTPRoute/TCPRoute` ' Defines the actual routes to Kubernetes services and their policies, such as timeouts, retries, and so on.

Here is an example GatewayClass resource:

---

gateway_class.yaml

```
apiVersion: gateway.networking.k8s.io/v1
kind: GatewayClass
metadata:
    name: external-lb
spec:
    controllerName: lb.acme.io/gateway-controller
```

This defines a GatewayClass named `external-lb` that will be handled by a `lb.acme.io/gateway-controller` controller.

A Gateway resource binds a hostname and TLS certificate to the GatewayClass as we can see in the following code:

---

gateway.yaml

```
apiVersion: gateway.networking.k8s.io/v1
kind: Gateway
metadata:
    name: my-gateway
spec:
    gatewayClassName: external-lb
    listeners:
    - name: http
    port: 80
    protocol: HTTP
```

This Gateway named `my-gateway` uses the `external-lb` GatewayClass defined earlier. It handles HTTP traffic on port `80`. Note that the `addresses` field is not specified, so an address or hostname will be assigned to the gateway by its controller.

Finally, HTTPRoute and TCPRoute resources define the actual routes to backend services. Here is an example:

http_route.yaml

```
apiVersion: gateway.networking.k8s.io/v1
kind: HTTPRoute
metadata:
    name: http-route
spec:
    parentRefs:
    - name: my-gateway
    rules:
    - matches:
    - path:
    type: PathPrefix
    value: /foo
    backendRefs:
    - name: my-foo-service
      port: 80
    - matches:
    - path:
        type: PathPrefix
        value: /bar
    backendRefs:
    - name: my-bar-service
      port: 80
```

This HTTPRoute is a child of the my-gateway Gateway defined earlier. It routes requests to the /foo path to the my-foo-service service on port 80 and requests to /bar are routed to my-bar-service on port 80. Also, additional features such as request timeouts, retries, and traffic splitting can be configured on the Route resources.

Gateways are a new and great way of configuring networking and routing in Kubernetes. The centralized configuration for ingress traffic management acts as a single source of truth. While the Ingress resource presents a simple, declarative interface focused specifically on exposing HTTP applications, the Gateway API resource offers a more generalized abstraction for proxying diverse protocols beyond HTTP. Also, they decouple the data plane from the control plane. Any gateway controller can be used, including NGINX, HAProxy, and Istio. Gateways provide improved security with TLS handling and authentication and fine-grained traffic control using advanced routing rules and policies. Finally, they have easier management and operation for complex ingress configurations.

Next, we will approach persistent volumes.

Persistent Volumes

Kubernetes was originally designed for stateless applications. So, one of the key challenges when running stateful applications on Kubernetes is managing storage. Kubernetes provides abstractions that allow storage to be provisioned and consumed in a portable manner across different environments. When designing storage infrastructure on Kubernetes, there are two main resources to understand: **PersistentVolumes** (**PVs**) and **PersistentVolumeClaims** (**PVCs**). A PV represents a networked storage unit provisioned by the cluster administrator. Much like compute nodes, PVs become a pool of cluster resources. In contrast, PVCs allow end users to request abstract storage with defined capacity and access modes. The PVC functions similarly to a pod resource request, but instead of CPU and memory, users can specify their desired volume size and read/write permissions. The Kubernetes control plane handles binding matching PV and PVC resources to provision storage for pods as declared. With this separation of roles, the underlying storage layer gains lifecycle independence from individual pods.

Here is an example PersistentVolume YAML manifest:

persistent_volume.yaml

```
apiVersion: v1
kind: PersistentVolume
metadata:
    name: pv0003
spec:
    capacity:
    storage: 5Gi
    volumeMode: Filesystem
    accessModes:
    - ReadWriteOnce
    persistentVolumeReclaimPolicy: Recycle
    storageClassName: slow
    mountOptions:
    - hard
    - nfsvers=4.1
    nfs:
    path: /tmp
    server: 172.17.0.2
```

This defines an **Network File System** (**NFS**) PV that supports the ReadWriteOnce access mode, has a capacity of 5GB, and is mounted at /tmp on the NFS server at 172.17.0.2. The reclaim policy is set to Recycle, which means the volume will be recycled rather than deleted when released. storageClassName is set to slow, which can be used to match this PV to PVCs requesting specific storage classes.

Here is an example PersistentVolumeClaim YAML manifest:

pvc.yaml

```
apiVersion: v1
kind: PersistentVolumeClaim
metadata:
    name: myclaim
spec:
    accessModes:
    - ReadWriteOnce
    volumeMode: Filesystem
    resources:
    requests:
      storage: 8Gi
    storageClassName: slow
    selector:
    matchLabels:
      release: "stable"
```

This PVC requests 8GB of storage with `ReadWriteOnce` access. It specifies the `slow` `storageClassName`, which will match it to the preceding PV with the same storage class. There is also a selector that will match PVs with a `stable` release label.

When a PVC is created, the Kubernetes control plane looks for a matching PV to bind to the PVC. This matching takes into account access modes, storage capacity, and `storageClassName` among other factors. Once bound, that storage is then available to be mounted by pods.

Here is a pod YAML that mounts the preceding PVC:

pod_with_pvc.yaml

```
kind: Pod
apiVersion: v1
metadata:
    name: mypod
spec:
    containers:
    - name: myfrontend
      image: nginx
      volumeMounts:
      - mountPath: "/var/www/html"
        name: mypd
    volumes:
```

```
  - name: mypd
    persistentVolumeClaim:
      claimName: myclaim
```

This pod mounts the PVC into the container at /var/www/html. The PVC provides durable storage for the pod that persists even if the pod is deleted or moved to a different node.

StorageClasses

The StorageClass objects define different *classes* of storage that can be requested. This allows administrators to offer different tiers of storage within the same cluster. The following code defines a standard hard disk StorageClass and a fast SSD StorageClass on GCE:

storage_classes.yaml

```
apiVersion: storage.k8s.io/v1
kind: StorageClass
metadata:
    name: standard
provisioner: kubernetes.io/gce-pd
parameters:
    type: pd-standard


---


apiVersion: storage.k8s.io/v1
kind: StorageClass
metadata:
    name: fast
provisioner: kubernetes.io/gce-pd
parameters:
    type: pd-ssd
```

The - - - line tells Kubernetes that we aggregated two YAML manifests in just one file. After defining a StorageClass, PVCs can then request a particular class:

pvc2.yaml

```
apiVersion: v1
kind: PersistentVolumeClaim
metadata:
    name: myclaim
spec:
    accessModes:
    - ReadWriteOnce
```

```
    storageClassName: fast
    resources:
    requests:
      storage: 30Gi
```

This allows a cluster to provide different types of storage without requiring users to understand how the details are implemented. Next, we will discuss the final subject in this chapter and one that is very important for security in Kubernetes: ConfigMaps and Secrets.

ConfigMaps and Secrets

ConfigMaps and Secrets are two important Kubernetes objects that allow you to separate configuration data from your application code. This makes your applications more portable, manageable, and secure.

ConfigMaps

ConfigMaps provide a convenient way to pass configuration data into pods in a declarative manner. They allow you to store configuration information without putting them directly in a pod definition or container image. Pods can access the data stored in a ConfigMap through environment variables, command-line arguments, or by mounting the ConfigMap as a volume. Using ConfigMaps enables you to separate your configuration data from your application code.

With the following manifest, you can create a ConfigMap to store configuration files:

config_map.yaml

```
apiVersion: v1
kind: ConfigMap
metadata:
    name: app-config
data:
    config.properties: |
    app.color=blue
    app.mode=prod
```

This ConfigMap contains a `config.properties` file that pods can mount and consume.

You can also create ConfigMaps from directories, files, or literal values. The following commands show an example of each ConfigMap definition. Those commands are run in a shell using the `kubectl` executable (we will take a deeper look at it in the next chapter):

```
kubectl create configmap app-config --from-file=path/to/dir

kubectl create configmap app-config --from-file=config.properties

kubectl create configmap app-config --from-literal=app.color=blue
```

To consume a ConfigMap in a pod, you can do the following:

- Set environment variables from ConfigMap data

- Set command-line arguments from ConfigMap data

- Consume ConfigMap values in volumes

This following YAML file defines a Kubernetes Pod that consumes configuration data from a ConfigMap using environment variables and consuming it as a volume. Let's see how that is done:

pod_with_configmap.yaml

```
apiVersion: v1
kind: Pod
metadata:
    name: configmap-demo
spec:
    containers:
    - name: demo
      image: alpine
      env:
        - name: APP_COLOR
          valueFrom:
            configMapKeyRef:
              name: app-config
              key: app.color
      args:
        - $(APP_MODE)
        valueFrom:
          configMapKeyRef:
            name: app-config
            key: app.mode
      volumeMounts:
        - name: config-volume
          mountPath: /etc/config
    volumes:
    - name: config-volume
      configMap:
        name: app-config
```

Let's take a closer look at this code and understand what it's doing:

- It defines a Pod with the name configmap-demo.

- The Pod has one container called demo that uses the alpine image.

- The container has two environment variables set:

    - `APP_COLOR` is set from the `app.color` key in the `app-config` ConfigMap

    - `APP_MODE` is set from the `app.mode` key in the `app-config` ConfigMap (this is defined as an argument to the run command)

- The container has one volume mount called `config-volume` that mounts the `/etc/config` path.

- The Pod defines a volume called `config-volume` that uses the `app-config` ConfigMap as a data source. This makes the data from the ConfigMap available to the container on the mount path.

Although ConfigMaps are really useful, they don't provide secrecy for confidential data. For that, Kubernetes provides Secrets.

Secrets

A Secret is an object that contains a small amount of sensitive data such as passwords, tokens, or keys. Secrets allow you to store and manage this sensitive data without exposing it in your application code.

For example, you can create a Secret from literal values using `kubectl`:

```
kubectl create secret generic db-secret \
--from-literal=DB_HOST=mysql \
--from-literal=DB_USER=root \
--from-literal=DB_PASSWORD=password123
```

The preceding code would create a Secret containing confidential database credentials. You can also create Secrets from files or directories:

```
kubectl create secret generic ssh-key-secret --from-file=ssh-
privatekey=/path/to/key

kubectl create secret generic app-secret --from-file=/path/to/dir
```

Secrets store data encoded in `base64` format. This prevents the values from being exposed as `plaintext` in `etcd`. However, the data is not encrypted. You can consume your secret data from pods like this:

pod_with_secrets.yaml

```
apiVersion: v1
kind: Pod
metadata:
```

```
     name: secret-demo
spec:
    containers:
    - name: demo
    image: nginx
    env:
      - name: DB_HOST
        valueFrom:
          secretKeyRef:
            name: db-secret
            key: DB_HOST
      - name: DB_USER
        valueFrom:
          secretKeyRef:
            name: db-secret
            key: DB_USER
      - name: DB_PASSWORD
        valueFrom:
          secretKeyRef:
            name: db-secret
            key: DB_PASSWORD
    volumeMounts:
      - name: secrets-volume
        mountPath: /etc/secrets
        readOnly: true
    volumes:
    - name: secrets-volume
      secret:
        secretName: app-secret
```

The preceding YAML file defines a Kubernetes Pod that consumes secrets from the Kubernetes API. Let's go through the code:

- It defines a Pod named secret-demo.

- The Pod has one container named demo based on the NGINX image.

- The container has three environment variables that get their values from secrets:

  - DB_HOST gets its value from the DB_HOST key in the db-secret secret

  - DB_USER gets its value from the DB_USER key in the db-secret secret

  - DB_PASSWORD gets its value from the DB_PASSWORD key in the db-secret secret

- The container mounts a volume called `secrets-volume` at the `/etc/secrets` path. This volume is read-only.

- The `secrets-volume` volume uses the `app-secret` secret as its backing store. So, any keys/values in `app-secret` will be exposed as files in `/etc/secrets` in the container.

Summary

In this chapter, we covered the fundamental architecture and components that make up a Kubernetes cluster. Understanding the control plane, nodes, and their components is crucial for operating Kubernetes effectively.

We looked at how the API server, etcd, controller manager, and schedulers in the control plane manage and maintain desired cluster state. kubelet and kube-proxy run on nodes to communicate with the control plane and manage containers. Getting familiar with these building blocks provides a mental model for how Kubernetes functions internally.

We also explored the main API resources used to deploy and manage applications, including Pods, Deployments, StatefulSets, Jobs, and Services. Pods encapsulate containers and provide networking and storage for closely related containers. Deployments and StatefulSets allow declarative management of pod replicas and provide self-healing capabilities. Jobs enable batch workloads to run to completion. Services enable loose coupling between pods and provide stable networking.

We discussed how ingress resources and ingress controllers configure external access to cluster services through routing rules. The new Gateway API provides a centralized way to manage ingress configuration. PersistentVolumes and PersistentVolumeClaims allow portable, network-attached storage to be provisioned and consumed efficiently. StorageClasses enable different classes of storage to be offered.

Finally, we looked at how ConfigMaps and Secrets allow configuration data and sensitive data to be injected into pods in a decoupled manner. Overall, these API resources provide powerful abstractions for deploying and managing applications robustly.

Learning these fundamental concepts equips you to use Kubernetes effectively. You now understand how a cluster is put together, how applications can be deployed and managed in line with the desired state, and how the supporting resources including storage, configuration, and secrets work. This critical foundation enables you to start deploying applications on Kubernetes and leverage its automation capabilities for self-healing, scaling, and management. The knowledge gained in this chapter will be indispensable as we move forward.

In the next chapter, we will do some hands-on exercises with Kubernetes to apply all the concepts that we studied here.

3

Getting Hands-On
with Kubernetes

In this chapter, we will get hands-on experience with Kubernetes by deploying both a local and cloud-based Kubernetes cluster, and then deploying sample applications into those clusters. First, you will deploy a local cluster using **Kubernetes in Docker** (**Kind**). Then, you will deploy managed Kubernetes clusters on AWS, GCP, and Azure. For the cloud options, we will provide the minimal account setup required to deploy the clusters. Feel free to choose the cloud provider you are most comfortable with; the core Kubernetes functionality will be the same.

After deploying a cluster, this chapter will be divided into two parts. In the first part, you will take the simple API application you developed in *Chapter 1* and deploy it into your Kubernetes cluster. You will learn how to containerize applications and work with Kubernetes deployments and services to expose your application. In the second part, you will deploy the simple data processing batch job from *Chapter 1* into Kubernetes. This will demonstrate how to run one-off jobs using Kubernetes.

By the end of this chapter, you will have first-hand experience deploying applications into Kubernetes. You will understand how to package and deploy containerized applications, expose them via services and ingress, and leverage Kubernetes for running both long-running services and batch jobs. With these skills, you will be ready to deploy applications into production Kubernetes environments.

In this chapter, we're going to cover the following main topics:

- Installing `kubectl`
- Deploying a local K8s cluster with Kind
- Deploying an AWS EKS cluster
- Deploying a Google Cloud GKE cluster
- Deploying an Azure AKS cluster
- Running your API on Kubernetes
- Running a data processing job in Kubernetes

Let's get to it.

Technical requirements

In this chapter, you will be required to have Docker installed (the instructions for this can be found in *Chapter 1*). The main hands-on activities we are going to do will be cloud-based, but you can choose to do them locally with Kind. You will learn how to deploy a Kubernetes cluster locally using Kind or in the cloud (AWS, Google Cloud, and Azure) in this chapter. Finally, you will need `kubectl` to interact with your Kubernetes cluster. You'll learn how to install it in the next section.

Installing kubectl

`kubectl` is a CLI that we are going to use to send commands to a Kubernetes cluster. You must have this installed so that you can interact with the cluster (regardless of whether it's running locally or in the cloud):

- To install `kubectl` on macOS with Homebrew, use the following command:

    ```
    brew install kubectl
    ```

- To install `kubectl` in a Linux distribution, you can use `curl` to download the binary executable:

    ```
    curl -LO "https://dl.k8s.io/release/$(curl -L -s https://dl.k8s.
    io/release/stable.txt)/bin/linux/amd64/kubectl"
    ```

- To install it in a Windows system, you can use the `chocolatey` package manager:

    ```
    choco install kubernetes-cli
    ```

From this point on, every `kubectl` command will be the same, regardless of the OS you're using. To check if the installation went successfully, run the following command. This will give you a nice, formatted view of the version of `kubectl` that's running on your system:

```
kubectl version –client –output=yaml
```

Now, let's move and install `kind`.

Deploying a local cluster using Kind

Deploying a local Kubernetes cluster can be extremely helpful for learning, testing, and preparing deployments for a production environment as Kubernetes is the same, wherever you run it. Let's start by deploying a single-node local cluster using Kind.

Installing kind

Kind is a tool that allows you to run Kubernetes on your local machine inside Docker containers. Besides being light and easy to set up, Kind delivers performance with the same Kubernetes standards. Kind clusters pass upstream Kubernetes conformance testing.

Kind is distributed as a single binary file. You can install it easily with package managers (make sure you have Docker already installed):

- If you're using macOS, use Homebrew to install `kind`, like so:

  ```
  brew install kind
  ```

- If you're using a Linux distribution (Ubuntu, for instance), you can install it with `curl`:

  ```
  $ [ $(uname -m) = x86_64 ] && curl -Lo ./kind https://kind.sigs.
  k8s.io/dl/v0.20.0/kind-linux-amd64
  $ chmod +x ./kind
  $ sudo mv ./kind /usr/local/bin/kind
  ```

- If you're using a Windows system, install it with `chocolatey` (`https://chocolatey.org/packages/kind`):

  ```
  choco install kind
  ```

After the installation is finished, you can check if it was installed correctly by running the following command:

```
kind version
```

Deploying the cluster

Now, to deploy a single-node local cluster, just run the following command:

```
kind create cluster
```

If this is the first time you've run this command, Kind will download the control plane image, create a Docker container for it, and configure a single-node Kubernetes cluster for you. The first time it runs, Kind will take 1-2 minutes to complete as it downloads the Kubernetes image. The next runs will be much faster.

To verify that the cluster is up, run the following command:

```
kubectl cluster-info
```

This will print connection details for the local cluster. And, that's it! You're good to go.

If you're willing to work with a cloud-based Kubernetes cluster, in the next few sections, we will deploy a cluster on AWS, Google Cloud, and Azure.

Deploying an AWS EKS cluster

Amazon Elastic Kubernetes Service (EKS) allows you to easily deploy and manage Kubernetes clusters on AWS. EKS handles provisioning and maintaining the Kubernetes control plane, while you can use standard Kubernetes tooling such as `kubectl` to manage worker nodes.

To get started with EKS, you need an AWS account. Go to `http://aws.amazon.com` and click **Create an AWS account**. Follow the steps to sign up for a new account; note that you will be requested to provide your credit card information. AWS offers a free usage tier that provides limited resources at no charge for 12 months. This is usually sufficient to run small workloads but not Kubernetes (although costs will not be high if you manage it wisely). AWS charges 73 dollars per month for every running cluster (assuming that the cluster is running for the whole month; if it runs just for a couple of days or hours, billing should be a fraction of that accordingly) plus the proper charging for each node that is running according to the size of the chosen machines.

After setting up the account, you must access it with your `root` user. We need to create an IAM user with specific permissions as this is the recommended usage. In the AWS console, go to the **IAM** service and click **Users** in the left panel. You should see a screen like this:

Figure 3.1 – The Users menu in IAM

Then, define the username and password for access. When you click **Next**, choose to attach policies directly (since this is only a study account – this configuration is not suited for a production environment). Choose **AdministratorAccess** from the list (which will permit you to do everything in AWS) and click **Next** to review. Then, click the **Create user** button to finish the process. When all is set, remember to download the AWS secrets (AWS access key ID and AWS secret access key). You will need these secrets to authenticate in AWS from your computer. Now, log off the console and access it again with your new IAM user to validate that your new IAM user has been configured correctly.

Before we look at the tools for setting up a Kubernetes cluster, we need to install the AWS CLI to interact with AWS from our local terminal. To do so, type the following:

```
pip3 install awscli
```

After installed, now we run aws configure to set up our AWS credentials:

```
aws configure
```

You will be prompted for your AWS credentials (AWS Access Key ID and AWS Secret Access Key). Copy and paste those credentials as you are asked for them. Then, the configuration will ask for the default AWS region. You can work in us-east-1. For the output format, json is a good choice.

This will save a config file with your credentials at ~/.aws/credentials. Now, you can run AWS CLI commands to interact with AWS services.

Once you have an AWS account and your IAM user and the AWS CLI have been configured, install the eksctl command-line tool. This tool simplifies and automates the process of deploying EKS clusters. Go to https://eksctl.io/installation/ and follow the installation instructions. Note that the documentation page lists the permissions you should have to deploy a cluster with eksctl. The **AdministratorAccess** policy should englobe any of the permissions you need.

After the installation, to check if eksctl has been set up correctly, run the following command:

```
eksctl version
```

You should see some output showing the version. Now, let's create a cluster from scratch with a single command:

```
eksctl create cluster \
    --managed --alb-ingress-access \
    --node-private-networking --full-ecr-access \
    --name=studycluster \
    --instance-types=m6i.xlarge \
    --region=us-east-1 \
    --nodes-min=2 --nodes-max=4 \
    --nodegroup-name=ng-studycluster
```

The base of this command is eksctl create cluster. From the second line on, we are stating some options. Let's look at them:

- The --managed option tells eksctl to create a fully managed cluster. It will handle creating the EKS control plane, node groups, networking, and more.

- The --alb-ingress-access option will configure the cluster to allow inbound traffic from load balancers. This is required for load balancer type services.

- The `--node-private-networking` option enables private networking between worker nodes. Nodes will have private IP addresses only.

- The `--full-ecr-access` option gives worker nodes full access to ECR container image repositories. This is needed to pull container images.

- The `--name` option sets a user-defined name for the cluster.

- The `--instance-type` option configures the instance type that will be used for worker nodes. In this case, we are choosing to use the `m6i.xlarge` EC2 instance.

- The `--region` option sets the AWS region – in this case, N. Virginia (us-east-1).

- The `--nodes-min` and `--nodes-max` options set an autoscaling rule for the node group. Here, we set the minimum to 2 instances and the maximum to 4 instances.

- The `--nodegroup` option sets the node group's name to `ng-studycluster`.

This will take around 10-15 minutes to complete. `eksctl` handles all the details of creating VPCs, subnets, security groups, IAM policies, and other AWS resources needed to run an EKS cluster.

Under the hood, `eksctl` uses CloudFormation to provision the AWS infrastructure. The `eksctl` command generates a CloudFormation template based on the parameters provided. It then deploys this template to create the necessary resources.

The following are some key components that are created by `eksctl`:

- **Virtual private cloud (VPC)**: A VPC network where your cluster resources will run. This includes public and private subnets across multiple availability zones

- **EKS cluster**: The Kubernetes control plane, which consists of the API server, etcd, controller manager, scheduler, and more. AWS operates and manages this

- **Worker node groups**: Managed node groups containing EC2 instances that will run your Kubernetes workloads

- **Security groups**: Firewall rules to control access to the cluster

- **IAM roles and policies**: Access policies to allow worker nodes and Kubernetes to access AWS APIs and resources

Once the `eksctl` command completes, your EKS cluster will be ready to use. You can view details about the cluster by running the following command:

```
eksctl get cluster --name studycluster --region us-east-1
```

Next, we are going to create a Kubernetes cluster using Google Cloud.

Deploying a Google Cloud GKE cluster

To deploy a Kubernetes cluster on Google Cloud, we need to set up a Google Cloud account and install the gcloud **command-line interface (CLI)**. To do that, go to https://cloud.google.com/ and click on **Start free** to create a new account. Follow the instructions and, when your new account is created, navigate to the console at https://console.cloud.google.com/. This is where you can manage all your Google Cloud resources.

Before we can deploy a GKE cluster, we need to enable the necessary APIs. Click on the navigation menu icon in the top left and go to **APIs & Services**. Search for Kubernetes Engine API and click on it. Make sure it is enabled. It is also a good practice to enable **Cloud Resource Manager API** and **Cloud Billing API**. This will allow us to create and manage the cluster resources. Next, we need to install and initialize the gcloud CLI. This will allow us to manage Google Cloud resources from the command line.

The gcloud CLI can be installed on Linux, macOS, and Windows. In a Linux distribution, you can download the installation script like so:

```
wget https://dl.google.com/dl/cloudsdk/release/google-cloud-sdk.tar.gz
```

Then, extract the archive:

```
tar -xzf google-cloud-sdk.tar.gz
```

Next, run the install script:

```
./google-cloud-sdk/install.sh
```

For a macOS installation, you can use Homebrew:

```
brew install --cask google-cloud-sdk
```

> **Note**
>
> For Windows, you can download the gcloud installer from https://cloud.google.com/sdk/docs/install#windows.

Once the installation is done, you can check it by running the following command:

```
gcloud --version
```

Once you've installed it, you must initiate the CLI:

```
gcloud init
```

This will walk you through logging in with your Google account and configuring `gcloud`.

Now that `gcloud` is installed and initialized, we can deploy a Kubernetes cluster on GKE. First, choose a Google Cloud project to deploy the cluster under. You can create a new project and set it as active like so:

```
gcloud projects create [PROJECT_ID]
gcloud config set project [PROJECT_ID]
```

Then, you must set a compute zone to deploy the cluster in. Here, we will work in the `us-central1-a` zone:

```
gcloud config set compute/zone us-central1-a
```

Now, we can deploy a Kubernetes cluster by running the following command:

```
gcloud container clusters create studycluster --num-nodes=2
```

The `--num-nodes` parameter controls how many nodes you are going to be deployed. Note that this will take several minutes to complete as Google Cloud will set up all the cluster components, networking, and so on. Once deployed, gather credentials to interact with the cluster:

```
gcloud container clusters get-credentials studycluster
```

This will save credentials to your Kubernetes config file. Finally, you can verify that you can connect to the cluster by running the following command:

```
kubectl cluster-info
```

You should see information about the Kubernetes master and the cloud provider. And that's it! You now have a fully functional Kubernetes cluster running on Google Cloud.

Next, we will do the same with Microsoft Azure Cloud.

Deploying an Azure AKS cluster

In this section, we will walk through the steps to deploy a Kubernetes cluster using **Azure Kubernetes Service (AKS)**. To get started with AKS, you need an Azure account. First, go to `https://azure.com`, click **Try Azure for free**, and then click **Start free**. This will allow you to start a free trial account on Azure. You will need to provide some basic information, such as your email and phone number, to set up the account. Make sure you use a valid email as Azure will send a verification code to complete the signup process. Once your account has been created, you will be directed to the Azure portal. This is the main dashboard for managing all your Azure resources.

At this point, it's recommended to install the Azure CLI on your local machine. The Azure CLI allows you to manage Azure resources from the command line. Follow the instructions at `https://docs.microsoft.com/en-us/cli/azure/install-azure-cli` to install it on Linux, macOS, or Windows.

After installing the CLI, run `az login` and follow the prompts to authenticate with your account. This will allow you to run Azure commands from your terminal. With the Azure account set up and the CLI installed, you are ready to deploy an AKS cluster.

The first thing you must do is create a resource group. Resource groups in Azure allow you to logically group resources such as AKS clusters, storage accounts, virtual networks, and more. Let's start by creating a resource group for our AKS cluster:

```
az group create --name myResourceGroup --location eastus
```

This will create a resource group named `myResourceGroup` in the `eastus` region. You can specify any region close to you. Now, we can create an AKS cluster in this resource group. The basic command to create a cluster is as follows:

```
az aks create --resource-group myResourceGroup --name studycluster--
node-count 2 --generate-ssh-keys
```

This will create a Kubernetes cluster named `studycluster` with two nodes. We generate SSH keys automatically to set up access to the nodes. Some other options you can specify are as follows:

- `--node-vm-size`: The size of the virtual machines for nodes. The default is `Standard_D2s_v3`.
- `--kubernetes-version`: The Kubernetes version to use for the cluster. It defaults to the latest version.
- `--enable-addons`: Enables add-ons such as monitoring, virtual nodes, and more.

The `az aks create` command handles setting up the Kubernetes cluster, virtual machines, networking, storage, and more automatically. The process may take 5-10 minutes to complete.

When it is done, you can connect to the cluster by running the following command:

```
kubectl get nodes
```

This will display the nodes that are part of your AKS cluster. At this point, you have successfully deployed an AKS cluster and are ready to deploy our API and batch processing job on it.

Running your API on Kubernetes

From this point, you can choose the Kubernetes deployment type you like the most (local or cloud-based) to run our API. The following examples will be shown with AWS but you also can choose another cloud provider. Feel free to do so.

Now, it is time to retake the API we built in *Chapter 1* and ship it to production. We developed a simple API that, when requested, can say hello to you or answer with a (very cool) joke.

We already have the code for the API (`https://github.com/PacktPublishing/Bigdata-on-Kubernetes/blob/main/Chapter01/app/main.py`) and the Dockerfile to build the container image (`https://github.com/PacktPublishing/Bigdata-on-Kubernetes/blob/main/Chapter01/Dockerfile`).

For the image to be accessible to Kubernetes, it should be available on a container registry. Each cloud provider has a registry but to make things simple, we are working with DockerHub (`https://hub.docker.com/`). So long as your images are public, you can store as many images as you want for free. Let's get started:

1. To get started, in your terminal, type the following command:

    ```
    docker login
    ```

    This will prompt you for your Docker username and password. If you don't have a DockerHub account yet, it will give you a website to create one. After logging in, we must build the image locally, like so:

    ```
    docker build -t <USERNAME>/jokeapi:v1 .
    ```

    Remember to replace `<USERNAME>` with your real DockerHub username. We are changing the image name to `jokeapi` so that it's easier to locate. If you're running Docker on a Mac M1, it is important to set the `--platform` parameter to make the container image compatible with AMD64 machines. To do so, run the following command:

    ```
    docker build --platform linux/amd64 -t <USERNAME>/jokeapi:v1 .
    ```

2. Now, we can push the image to DockerHub:

    ```
    docker push <USERNAME>/jokeapi:v1
    ```

3. Go to `https://hub.docker.com/` and log in. You should see your image listed in the **Repositories** section.

Now, the image is available for Kubernetes. Next, we need to define some Kubernetes resources to run our API. We will create a deployment and a service.

Creating the deployment

Follow these steps:

1. First, we'll create the deployment. This specifies how many Pod replicas to run and their configuration:

    deployment_api.yaml

    ```yaml
    apiVersion: apps/v1
    kind: Deployment
    metadata:
        name: jokeapi
    spec:
        replicas: 2
        selector:
        matchLabels:
          app: jokeapi
        template:
        metadata:
          labels:
            app: jokeapi
        spec:
          containers:
          - name: jokeapi
            image: <USERNAME>/jokeapi:v1
            imagePullPolicy: Always
            ports:
            - containerPort: 8087
    ```

    This will run two Pod replicas using the Docker image we built. Note that we are opening port `8087` in the container. This is similar to the `EXPOSE` command in the Dockerfile.

2. Next, we will create a namespace to separate and organize our resources and apply the deployment:

    ```
    kubectl create namespace jokeapi
    kubectl apply -f deployment_api.yaml -n jokeapi
    ```

3. This will create the deployment and the two Pod replicas in the `jokeapi` namespace. We can check that everything is working by running the following command:

    ```
    kubectl get deployments -n jokeapi
    ```

    You should see the following output:

    ```
    NAME       READY    UP-TO-DATE    AVAILABLE    AGE
    jokeapi    0/2      2             0            2s
    ```

4. Now, let's check if the Pods are running properly:

```
kubectl get pods -n jokeapi
```

You should see an output like this:

```
NAME                      READY   STATUS    RESTARTS
jokeapi-7d9877598d-bsj5b   1/1     Running   0
jokeapi-7d9877598d-qb8vs   1/1     Running   0
```

Creating a service

Follow these steps:

1. Here, we'll specify a service to expose the Pods in the cluster:

service_api.yaml

```
apiVersion: v1
kind: Service
metadata:
    name: jokeapi
spec:
    selector:
    app: jokeapi
    ports:
    - protocol: TCP
      port: 80
      targetPort: 8087
```

This creates a ClusterIP service that exposes the API Pods on an internal IP address in the cluster. Note that the `type` parameter is not specified within the `spec` section of the YAML file, so, it defaults to ClusterIP.

2. To make the API accessible externally, we can create a load balancer (not possible with a local `kind` cluster, only with cloud-based clusters):

lb_api.yaml

```
apiVersion: v1
kind: Service
metadata:
    name: jokeapi
spec:
    selector:
    app: jokeapi
    type: LoadBalancer
    ports:
```

```
     - protocol: TCP
       port: 80
       targetPort: 8087
```

3. Now, the `type` definition is set and this code will define a load balancer and assign an external IP. Next, we will deploy the load balancer service:

    ```
    kubectl apply -f lb_api.yaml -n jokeapi
    ```

4. Now, we can test if the API is accessible. First, we must get the load balancer's URL (here, we are working in AWS) by running the following command:

    ```
    kubectl get services -n jokeapi
    ```

    You should see the following output:

    ```
    NAME       TYPE          CLUSTER-IP       EXTERNAL-IP
    jokeapi    LoadBalancer  10.100.251.249   <DNS>.amazonaws.com
    ```

5. Copy the content under `EXTERNAL-IP`, paste the URL in a browser, and add `/joke`. For instance, in my implementation here, I got `ab1cdd20ce1a349bab9af992211be654-1566834308.us-east-1.elb.amazonaws.com/joke`. You should see the following response on your screen:

    ```
    {"joke":"Have you heard about the chocolate record player? It
    sounds pretty sweet."}
    ```

Success! We have a (great) joke in our browser! Now, we will deploy the API with an ingress instead of a load balancer (for cloud-based clusters only).

Using an ingress to access the API

For this deployment, we will use the NGINX ingress controller and connect it to the load Balancer provided by AWS (the process is very similar if you are working with any cloud provider). Follow these steps:

1. First, we'll create a new namespace for NGINX and deploy the controller on Kubernetes:

    ```
    kubectl create namespace ingress-nginx
    kubectl apply -f https://raw.githubusercontent.com/kubernetes/
    ingress-nginx/controller-v1.1.3/deploy/static/provider/
    baremetal/deploy.yaml -n ingress-nginx
    ```

2. This will deploy the NGINX controller using the official manifests. Now, we have to edit one line in the deployment to make sure it uses the load balancer as an ingress deployment and not `NodePort`, its default. In your terminal, type the following:

    ```
    kubectl edit service ingress-nginx-controller -n ingress-nginx
    ```

3. Search for the `spec.type` field and change its value to `LoadBalancer`. After saving the file, let's check the services that have been deployed with the controller:

    ```
    kubectl get services -n ingress-nginx
    ```

4. We will see that the `ingress-nginx-controller` service is set to `LoadBalancer` and has an external IP related to it. Now, it is easy to set up an ingress that points to this ingress controller. First, we'll create a service defined in the `service_api.yaml` file. This service should be set to a ClusterIP type (see the code in the previous section). Then, we can define an ingress with the following code:

    ingress.yaml

    ```yaml
    apiVersion: networking.k8s.io/v1
    kind: Ingress
    metadata:
        name: jokeapi-ingress
    spec:
        rules:
        - http:
          paths:
          - path: /
            pathType: Prefix
            backend:
              service:
                name: jokeapi
                port:
                  number: 80
    ```

5. This ingress will route external traffic to the internal service IP. Once the ingress has an external IP assigned, we should be able to access our API by hitting that URL. Type the following

    ```
    kubectl get services -n ingress-nginx
    ```

6. Get the external URL for the controller and add `/joke` to it:

    ```
    {"joke":"What do you call a fish wearing a bowtie?
    Sofishticated."}
    ```

Et voilà! In the next section, we will deploy our data processing job on Kubernetes.

Running a data processing job in Kubernetes

In this section, we will deploy the simple data processing job from *Chapter 1* on Kubernetes. We have already developed the job (`https://github.com/PacktPublishing/Bigdata-on-Kubernetes/blob/main/Chapter01/run.py`) and built a Dockerfile to package it into a container image (`https://github.com/PacktPublishing/Bigdata-on-Kubernetes/blob/main/Chapter01/Dockerfile_job`).

Now, we have to build a Docker image and push it to a repository that's accessible to Kubernetes.

```
docker build --platform linux/amd64 -f Dockerfile_job -t <USERNAME>/
dataprocessingjob:v1 .
docker push <USERNAME>/dataprocessingjob:v1
```

Now, we can create a Kubernetes job to run our data processing task. Here's an example job manifest:

job.yaml

```
apiVersion: batch/v1
kind: Job
metadata:
    name: dataprocessingjob
spec:
    template:
    spec:
      containers:
      - name: dataprocessingjob
        image: <USERNAME>/dataprocessingjob:v1
      restartPolicy: Never
    backoffLimit: 4
```

This configures a job named dataprocessingjob that will run one replica of the <USERNAME>/
dataprocessingjob:v1 image. Now, we can create a new namespace and deploy the job, like so:

```
kubectl create namespace datajob
kubectl apply -f job.yaml -n datajob
```

This defines a job called dataprocessingjob that will run a single Pod using our Docker image. We
set restartPolicy: Never since we want the container to run to completion rather than restart.

We can check the status of the job like so:

```
kubectl get jobs -n datajob
```

After the job has been completed, we will see 1/1:

```
NAME                  COMPLETIONS    DURATION    AGE
dataprocessingjob        1/1            8s        11s
```

To view the logs from our job, we can use kubectl logs on the Pod created by the job:

```
kubectl get pods -n datajob
kubectl logs <NAMEOFTHEPOD> -n datajob
```

In my case, I typed the following:

```
kubectl logs dataprocessingjob-g8lkm -n datajob
```

I got the following results:

```
Index([0, 1, 2, 3, 4, 5, 6, 7, 8, 'newcolumn'], dtype='object')
     0    1    2    3      4      5      6    7   8    newcolumn
0    6  148   72   35      0   33.6  0.627   50   1       67.2
1    1   85   66   29      0   26.6  0.351   31   0       53.2
2    8  183   64    0      0   23.3  0.672   32   1       46.6
3    1   89   66   23     94   28.1  0.167   21   0       56.2
4    0  137   40   35    168   43.1  2.288   33   1       86.2
(768, 10)
```

This will print the application output from our Python program so that we can verify it ran correctly. And that's it! You ran a data processing job inside Kubernetes!

Summary

In this chapter, we gained hands-on experience deploying Kubernetes clusters and running applications in them. We started by installing `kubectl` and deploying a local Kubernetes cluster using Kind. Then, we deployed managed Kubernetes clusters on AWS, GCP, and Azure. While the cloud providers differ, Kubernetes provides a consistent environment to run containers.

After setting up our clusters, we containerized and deployed the simple API application from *Chapter 1*. This demonstrated how to define Kubernetes deployments, services, ingress, and load balancers to run web applications. Then, we deployed the data processing batch job from *Chapter 1* as a Kubernetes job. This showed us how to leverage Kubernetes for running one-off tasks and jobs.

By going through the process of deploying clusters and applications end-to-end, you now have first-hand experience with Kubernetes. You understand how to package applications as containers, expose them via services, ingress, or load balancers, and leverage Kubernetes abstractions such as deployments and jobs. With these skills, you are equipped to run applications and workloads on Kubernetes in development or production environments.

In the next chapter, we are going to take a closer look at the modern data stack, understanding each technology, why they are important, and how they link together to build a data solution.

Part 2:
Big Data Stack

In this part, you will dive into the core technologies that make up the **modern data stack**, a set of tools and architectures designed for building robust and scalable data pipelines. You will gain a solid understanding of the Lambda architecture and its components, and gain some hands-on experience with powerful big data tools such as Apache Spark, Apache Airflow, and Apache Kafka.

This part contains the following chapters:

- *Chapter 4, The Modern Data Stack*
- *Chapter 5, Big Data Processing with Apache Spark*
- *Chapter 6, Apache Airflow for Building Pipelines*
- *Chapter 7, Apache Kafka for Real-Time Events and Data Ingestion*

4

The Modern Data Stack

In this chapter, we will explore the modern data architecture that has emerged for building scalable and flexible data platforms. Specifically, we will cover the Lambda architecture pattern and how it enables real-time data processing along with batch data analytics. You will learn about the key components of the Lambda architecture, including the batch processing layer for historical data, the speed processing layer for real-time data, and the serving layer for unified queries. We will discuss how technologies such as Apache Spark, Apache Kafka, and Apache Airflow can be used to implement these layers at scale.

By the end of the chapter, you will understand the core design principles and technology choices for building a modern data lake. You will be able to explain the benefits of the Lambda architecture over traditional data warehouse designs. Most importantly, you will have the conceptual foundation to start architecting your own modern data platform.

The concepts covered will allow you to process streaming data at low latency while also performing complex analytical workloads on historical data. You will gain practical knowledge on leveraging open source big data technologies to build scalable and flexible data pipelines. Whether you need real-time analytics, machine learning model training, or ad-hoc analysis, modern data stack patterns empower you to support diverse data needs.

This chapter provides the blueprint for transitioning from legacy data warehouses to next-generation data lakes. The lessons equip you with the key architectural principles, components, and technologies to build modern data platforms on Kubernetes.

In this chapter, we're going to cover the following main topics:

- Data architectures
- Data lake design for big data
- Implementing the lakehouse architecture

Data architectures

Modern data architectures have evolved significantly over the past decade to enable organizations to harness the power of big data and drive advanced analytics. Two key architectural patterns that have emerged are the Lambda and Kappa architectures. In this section, we will have a look at both of them and understand how they can provide a useful framework for structuring our big data environment.

The Lambda architecture

The Lambda architecture is a big data processing architecture pattern that balances batch and real-time processing methods. Its name comes from the Lambda calculus model of computation. The Lambda architecture became popular in the early 2010s as a way to handle large volumes of data in a cost-effective and flexible manner.

The core components of the Lambda architecture include the following:

- **Batch layer**: Responsible for managing the master dataset. This layer ingests and processes data in bulk at regular intervals, typically every 24 hours. Once processed, the batch views are considered immutable and stored.

- **Speed layer**: Responsible for recent data that has not yet been processed by the batch layer. This layer processes data in real time as it arrives to provide low-latency views.

- **Serving layer**: Responsible for responding to queries by merging views from both the batch and speed layers.

Those components are presented in the architecture diagram in *Figure 4.1*:

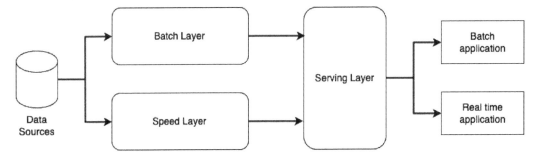

Figure 4.1 – Lambda architecture design

The key benefit of the Lambda architecture is that it provides a hybrid approach that combines historical views of large data volumes (batch layer) with up-to-date views of recent data (speed layer). This enables analysts to query both recent and historical data in a unified way to gain quick insights.

The batch layer is optimized for throughput and efficiency while the speed layer is optimized for low latency. By separating the responsibilities, the architecture avoids having to run large-scale, long-running batch jobs for every query. Instead, queries can leverage pre-computed batch views and augment them with up-to-date data from the speed layer.

In a modern data lake built on cloud infrastructure, the Lambda architecture provides a flexible blueprint. The cloud storage layer serves as the foundational data lake where data is landed. The batch layer leverages distributed data processing engines such as Apache Spark to produce batch views. The speed layer streams and processes the most recent data, and the serving layer runs performant query engines such as Trino to analyze data.

The Kappa architecture

The Kappa architecture emerged more recently as an alternative approach from primarily the same creators of the Lambda architecture. The main difference in the Kappa architecture is that it aims to simplify the Lambda model by eliminating the separate batch and speed layers.

Instead, the Kappa architecture handles all data processing through a single stream processing pathway. The key components include the following:

- **Stream processing layer**: Responsible for ingesting and processing all data as streams. This layer handles both historical data (via replay of logs/files) as well as new incoming data.

- **Serving layer**: Responsible for responding to queries by accessing views produced by the stream processing layer.

We can see a visual representation in *Figure 4.2*:

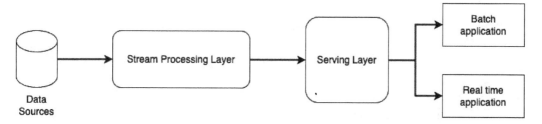

Figure 4.2 – Kappa architecture design

At the core of Kappa is an immutable, append-only log for all data-using tools such as Kafka and event-sourcing paradigms. Streaming data is ingested directly into the log instead of separate pipelines. The log ensures ordered, tamper-proof data with automatic replayability – key enablers for both stream and batch processing.

The benefit of the Kappa architecture is its design simplicity. By having a single processing pathway, there is no need to manage separate batch and real-time systems. All data is handled through stream processing, which also enables flexible reprocessing and analysis of historical data.

The trade-off is that stream processing engines may not offer the same scale and throughput as the most advanced batch engines (although modern stream processors have continued to evolve to handle very large workloads). Also, while Kappa design can be simpler, the architecture itself can be much harder to implement and maintain than Lambda.

For data lakes, the Kappa architecture aligns well with the nature of large volumes of landing data. The cloud storage layer serves as the raw data backbone. Then, stream processors such as Apache Kafka and Apache Flink ingest, process, and produce analysis-ready views of the data. The serving layer leverages technologies such as Elasticsearch and MongoDB to power analytics and dashboards.

Comparing Lambda and Kappa

The Lambda and Kappa architectures take different approaches but solve similar needs in preparing, processing, and analyzing large datasets. Key differences are listed in *Table 4.1*:

	Lambda	Kappa
Complexity	Manages separate batch and real-time systems	Consolidates processing through streams
Reprocessing	Reprocesses historical batches	Relies on stream replay and algorithms
Latency	Lower latencies for recent data in the speed layer	Same latency for all data
Throughput	Leverages batch engines optimized for throughput	Processes all data as streams

Table 4.1 – Lambda and Kappa architecture main differences

In practice, modern data architectures often blend these approaches. For example, a batch layer on Lambda may run only weekly or monthly while real-time streams fill the gap. Kappa may leverage small batches within its streams to optimize throughput. The core ideas around balancing latency, throughput, and reprocessing are shared.

For data lakes, Lambda provides a proven blueprint while Kappa offers a powerful alternative. While some may argue that Kappa offers a simpler operation, it is hard to implement and its costs can grow rapidly with scale. Another advantage of Lambda is that it is fully adaptable. We can implement only the batch layer if no data streaming is necessary (or financially viable).

Data lake builders should understand the key principles of each to craft optimal architectures aligned to their businesses, analytics, and operational needs. By leveraging the scale and agility of cloud infrastructure, modern data lakes can implement these patterns to handle today's data volumes and power advanced analytics.

In the next section, we will dive deeper into the Lambda architecture approach and how it can be applied to creating performant, scalable data lakes.

Data lake design for big data

In this section, we will contrast data lakes with traditional data warehouses and cover core design patterns. This will set the stage for the hands-on tools and implementation coverage in the final "How to" section. Let's start with the baseline for the modern data architecture: the data warehouse.

Data warehouses

Data warehouses have been the backbone of business intelligence and analytics for decades. A data warehouse is a repository of integrated data from multiple sources, organized and optimized for reporting and analysis.

The key aspects of the traditional data warehouse architecture are as follows:

- **Structured data**: Data warehouses typically only store structured data such as transaction data from databases and CRM systems. Unstructured data from documents, images, social media, and so on are not included.

- **Schema-on-write**: The data structure and schema are defined upfront during data warehouse design. This means adding new data sources and changing business requirements can be difficult.

- **Batch processing**: Data is **extracted, transformed, and loaded** (ETL) from source systems in batches according to a schedule, often daily or weekly. This introduces latency when accessing up-to-date data.

- **Separate from source systems**: The data warehouse acts as a separate store of data optimized for analytics, independent from the source transactional systems.

The growth of data volumes, variety, and velocity in the era of big data exposed some limitations with the traditional data warehouse architecture.

It could not cost-effectively store and process huge volumes of unstructured and semi-structured data from new sources such as websites, mobile apps, IoT devices, and social media. Also, it lacked flexibility – adding new data sources required changes to schemas and ETL, which made adaptations slow and expensive. Finally, batch processing couldn't deliver insights quickly enough for emerging requirements such as real-time personalization and fraud detection.

This gave rise to the data lake architecture in response, which we will see in detail next.

The rise of big data and data lakes

In response to the aforementioned challenges, the new data lake approach made it possible to deal with huge storage of any type of data at scale, using affordable distributed storage such as Hadoop HDFS or cloud object storage. Data lakes operate in a schema-on-read way instead of an upfront schema. Data is stored in native formats, and only the schema is interpreted at the time of reading. It includes the capture, storage, and access of real-time streaming data via tools such as Apache Kafka. There is also a big open source ecosystem for scalable processing including MapReduce, Spark, and other tools.

A data lake is a centralized data repository. It is designed to store data in its raw format as-is. This provides the flexibility to analyze different types of data on demand (tables, images, text, videos, etc.), instead of needing to predetermine how it will be used. Because it is implemented on top of object storages, it can store a huge amount of data coming from anywhere in different intervals (some data can come daily, some hourly, and some in near real time). Data lakes also separate storage and processing technology (different from the data warehouse, where storage and processing happen in a whole unique structure). Usually, data processing involves a distributed compute engine (such as Spark) for terabyte-scale processing.

Data lakes provided a way to cost-effectively store the huge and diverse data volumes that organizations were grappling with and perform analytics. However, they also had some challenges:

- Without governance, data lakes risked becoming inaccessible data *swamps*. Data needed to be cataloged with context for discoverability.

- Preparing raw data for analysis still involved complex data wrangling across disparate siloed tools.

- Most analytics still required data to be modeled, cleansed, and transformed first – such as a data warehouse. This duplicated efforts.

- The object-based storage systems used in data lakes did not allow to perform line-level modifications. Whenever a line in a table needed to be modified, the whole file would be rewritten, causing a big impact on processing performance.

- In data lakes, there is not an efficient schema control. While the schema-on-read approach makes it easier for new data sources, there is no guarantee that tables will not change their structure because of a failed ingestion.

In recent years, there has been a major effort to overcome these new challenges by joining the best of both worlds, which is now known as the data lakehouse. Let's dive into this concept.

The rise of the data lakehouse

In the 2010s, the term "lakehouse" gained attention because of new open source technologies such as Delta Lake, Apache Hudi, and Apache Iceberg. The lakehouse architecture aims to combine the best aspects of data warehouses and data lakes:

- Supporting diverse structured and unstructured data at any scale like a data lake

- Providing performant SQL analytics across raw and refined data like a data warehouse

- **ACID (atomic, consistent, isolated, and durable)** transactions on large datasets

Data lakehouses allow storing, updating, and querying data simultaneously in open formats while ensuring correctness and reliability at scale. This enables features such as the following:

- Schema enforcement, evolution, and management

- Line-level *upserts* (updates + inserts) and deletes for performant mutability

- Point-in-time consistency views across historic data

Using the lakehouse architecture, the entire analytics life cycle, from raw data to cleaned and modeled data to curated data products, is directly accessible in one place for both batch and real-time use cases. This drives greater agility, reduces duplication of efforts, and enables easier reuse and repurposing of data through the life cycle.

Next, we will look at how data is structured within this architecture concept.

The lakehouse storage layers

Like the data lake architecture, the lakehouse is also built on cloud object storage, and it is commonly divided into three main layers: bronze, silver, and gold. This approach became known as the "medallion" design.

The bronze layer is the raw ingested data layer. It contains the original raw data from various sources, stored exactly as it was received. The data formats can be structured, semi-structured, or unstructured. Examples include log files, CSV files, JSON documents, images, audio files, and so on.

The purpose of this layer is to store the data in its most complete and original format, acting as the version of truth for analytical purposes. No transformations or aggregations happen at this layer. It serves as the source for building curated and aggregated datasets in the higher layers.

The silver layer contains curated, refined, and standardized datasets that are enriched, cleaned, integrated, and conformed to business standards. The data has consistent schemas and is queryable for analytics.

The purpose of this layer is to prepare high-quality, analysis-ready datasets that can feed into downstream analytics and machine learning models. This involves data wrangling, standardization, deduplication, joining disparate data sources, and so on.

The structure can be tables, views, or files optimized for querying. Examples include Parquet, Delta Lake tables, materialized views, and so on. Metadata is added to enable data discovery.

The gold layer contains aggregated data models, metrics, KPIs, and other derivative datasets that power business intelligence and analytics dashboards.

The purpose of this layer is to serve ready-to-use curated data models to business users for reporting and visualization. This involves pre-computing metrics, aggregations, business logic, and so on to optimize for analytical workloads.

The structure optimizes analytics through columnar storage, indexing, partitioning, and so on. Examples include aggregates, cubes, dashboards, and ML models. Metadata ties this to upstream data.

Sometimes, it is common to have an extra layer – a landing zone before the bronze layer. In this case, the landing zone receives raw data, and all the cleansing and structuring are done in the bronze layer.

In the next section, we will see how to operationalize the data lakehouse design with modern data engineering tools.

Implementing the lakehouse architecture

Figure 4.3 shows a possible implementation of a data lakehouse architecture in a Lambda design. The diagram shows the common lakehouse layers and the technologies used to implement this on Kubernetes. The first group on the left represents the possible data sources to work with this architecture. One of the key advantages of this approach is its ability to ingest and store data from a wide variety of sources and in diverse formats. As shown in the diagram, the data lake can connect to and integrate structured data from databases as well as unstructured data such as API responses, images, videos, XML, and text files. This schema-on-read approach allows the raw data to be loaded quickly without needing upfront modeling, making the architecture highly scalable. When analysis is required, the lakehouse layer enables querying across all these datasets in one place using schema-on-query. This makes it simpler to integrate data from disparate sources to gain new insights. The separation of loading from analysis also enables iterative analytics as a new understanding of the data emerges. Overall, the modern data lakehouse is optimized for rapidly landing multi-structured and multi-sourced data while also empowering users to analyze it flexibly.

First, we will take a closer look at the batch layer shown at the top of *Figure 4.3*.

Batch ingestion

The first layer of the design refers to the batch ingestion process. For all the unstructured data, customized Python processes are the way to go. It is possible to develop custom code to query data from API endpoints, to read XML structures, and to process text and images. For structured data in databases, we have two options for data ingestion. First, Kafka and Kafka Connect provide a way of simply configuring data migration jobs and connecting to a large set of databases. Apache Kafka is a distributed streaming platform that allows publishing and subscribing to streams of records. At its core, Kafka is a durable message broker built on a publish-subscribe model. Kafka Connect is a tool included with Kafka that provides a generic way to move data into and out of Kafka. It offers reusable connectors that can help connect Kafka topics to external systems such as databases, key-value stores, search indexes, and filesystems. Kafka Connect features connector plugins for many common data sources and sinks such as JDBC, MongoDB, Elasticsearch, and so on. These connectors move data from the external system into Kafka topics and vice versa.

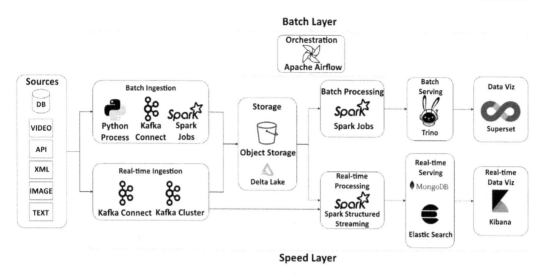

Figure 4.3 – Data lakehouse in Kubernetes

These connectors are reusable and configurable. For example, the JDBC connector can be configured to capture changes from a PostgreSQL database and write them to Kafka topics. Kafka Connect handles translating the data formats, distributed coordination, fault tolerance, and so on, and it supports stream processing by tracking data changes in the source connectors (e.g., database **change data capture** (**CDC**) connectors) and piping the change stream into Kafka topics. This simplifies the process of getting data in and out of Kafka. Although Kafka is a well-known tool for streaming data, its use alongside Kafka Connect has proven extremely efficient for batch data migration from databases.

Sometimes, when managing a Kafka cluster for data migration is not viable (we will talk about some of these cases later), it is possible to ingest data from structured sources with Apache Spark. Apache Spark provides a versatile tool for ingesting data from various structured data sources into a data lake built on cloud object storage such as Amazon S3 or Azure Data Lake Storage. The Spark DataFrame API allows querying data from relational databases, NoSQL data stores, and other structured data sources. While convenient, reading from JDBC data sources in Spark can be inefficient. Spark will read the table as a single partition, so all processing will occur in a single task. For large tables, this can slow down ingestion and subsequent querying (more details in *Chapter 5*). To optimize, we need to manually partition the reading from the source database. The main drawback with Spark data ingestion is handling these partitioning and optimization concerns yourself. Other tools can help by managing parallel ingestion jobs for you, but Spark gives the flexibility to connect and process many data sources out of the box.

Now, let's take a look at the storage layer.

Storage

Next, in the middle of the diagram, we have the **storage** layer. This is the only one I do not recommend moving to Kubernetes. Cloud-based object storage services now have plenty of features that optimize scalability and reliability, making it simple to operate and with great retrieval performance. Although there are some great tools for building a data lake storage layer in Kubernetes (e.g., `https://min.io/`), it is not worth the effort, since you would have to take care of scalability and reliability yourself. For the purposes of this book, we will work with all the lakehouse layers in Kubernetes except the storage layer.

Batch processing

Now, we will talk about the **batch processing** layer. Apache Spark has become the de facto standard for large-scale batch data processing in the big data ecosystem. Unlike traditional MapReduce jobs that write intermediate data to disk, Spark processes data in memory, making it much faster for iterative algorithms and interactive data analysis. Spark utilizes a cluster manager to coordinate job execution across a group of worker nodes. This allows it to process very large datasets by distributing the data across the cluster and parallelizing the processing. Spark can efficiently handle terabytes of data stored in distributed filesystems such as HDFS and cloud object stores.

One of the key advantages of Spark is the unified API it provides for both SQL and complex analytics. Data engineers and scientists can use the Python DataFrame API to process and analyze batch datasets. The same DataFrames can then be queried through Spark SQL, providing familiarity and interactivity. This makes Spark very simple to operate for a wide range of users. By leveraging in-memory processing and providing easy-to-use APIs, Apache Spark has become the go-to solution for scalable batch data analytics. Companies with large volumes of log files, sensor data, or other records can rely on Spark to efficiently process these huge datasets in parallel. This has cemented its place as a foundational technology in modern data architectures.

Next, we will discuss the orchestration layer.

Orchestration

Above the storage layer and the batch processing layer, in *Figure 4.3*, we find an **orchestration** layer. As we build more complex data pipelines that chain together multiple processing steps, we need a way to reliably manage the execution of these pipelines. This is where orchestration frameworks come in. Here, we chose to work with Airflow. Airflow is an open source workflow orchestration platform originally developed at Airbnb to author, schedule, and monitor data pipelines. It has since become one of the most popular orchestration tools for data pipelines.

The key reasons why using Airflow is important for batch data pipelines are as follows:

- **Scheduling**: Airflow allows you to schedule batch jobs to run periodically (hourly, daily, weekly, etc.). This removes the need to manually kick off jobs and ensures they run reliably.

- **Dependency management**: Jobs often need to run sequentially or wait for other jobs to complete. Airflow provides an easy way to set up these dependencies in a **directed acyclic graph (DAG)**.

- **Monitoring**: Airflow has a built-in dashboard to monitor the status of jobs. You get visibility of what has succeeded, failed, is currently running, and so on. It also keeps logs and history for later debugging.

- **Flexibility**: New data sources, transformations, and outputs can be added by modifying the DAG without impacting other non-related jobs. Airflow DAGs provide high configurability.

- **Abstraction**: Airflow DAGs allow pipeline developers to focus on the business logic rather than application orchestration. The underlying Airflow platform handles the workflow scheduling, status monitoring, and so on.

Now, we will move on to the serving layer.

Batch serving

For the **batch serving** layer in Kubernetes, we have chosen to work with Trino. Trino (formerly known as PrestoSQL) is an open source, distributed SQL query engine built for executing interactive analytic queries against a variety of data sources. Trino can be used to run queries up to a petabyte scale. With Trino, you can query multiple data sources in parallel. When a SQL query is submitted to Trino, it is parsed and planned to create a distributed execution plan. This execution plan is then submitted to worker nodes that process the query in parallel and return the results to the coordinator node. It supports ANSI SQL (one of the most common patterns for SQL) and it can connect to a variety of data sources, including all the main cloud-based object storage services. By leveraging Trino, data teams can enable self-service SQL analytics directly on their cloud data lakes. This eliminates costly and slow data movement just for analytics while still providing interactive response times.

Next, we will take a look at the tools chosen for data visualization.

Data visualization

For data visualization and analytics, we chose to work with Apache Superset. Although there are many great tools for this on the market, we find Superset easy to deploy, easy to run, easy to use, and extremely easy to integrate. Superset, an open source data exploration and visualization application, enables users to build interactive dashboards, charts, and graphs with ease. Superset originated at Airbnb in 2015 as an internal tool for its analysts and data scientists. As Airbnb's usage and contributions grew, it decided to open source Superset in 2016 under the Apache license and donated it to the Apache Software Foundation. Since then, Superset has been adopted by many other companies and has an active open source community contributing to its development. It has an intuitive graphical interface to visualize and explore data through rich dashboards, charts, and graphs that support many complex visualization types out of the box. It has a SQL Lab editor that allows you to write SQL queries against different databases and visualize results. It provides secure access and role management that allows granular control over data access and modification. It can connect to a great variety of data sources, including relational databases, data warehouses, and SQL engines such as Trino. Superset can be conveniently deployed on Kubernetes using Helm charts that are provided. The Helm chart provisions all the required Kubernetes objects – deployments, services, ingress, and so on, to run Superset.

With rich visualization capabilities, the flexibility to work with diverse data sources, and Kubernetes' deployment support, Apache Superset is a valuable addition to the modern data stack on Kubernetes.

Now, let's move on to the bottom part of the diagram in *Figure 4.3*, the real-time layer.

Real-time ingestion

In batch data ingestion, data is loaded in larger chunks or batches on a regular schedule. For example, batch jobs may run every hour, day, or week to load new data from source systems. On the other hand, in real-time data ingestion, data is streamed into the system continuously as it is generated. This enables a true, near-real-time flow of data into the data lake. Real-time data ingestion is *event-driven* – as events occur, they generate data that flows into the system. This could include things such as user clicks, IoT sensor readings, financial transactions, and so on. The system reacts to and processes each event as it arrives. Apache Kafka is one of the most popular open source tools that provide a scalable, fault-tolerant platform for handling real-time data streams. It can be used with Kafka Connect for streaming data from databases and other structured data sources or with customized data producers developed in Python, for instance.

Data ingested in real time on Kafka is usually also "synced" to the storage layer for later historical analysis and backup. It is not recommended that we use Kafka as our only real-time data storage. Instead, we apply the best practice of erasing data from it to save storage space after a defined period. The default period for this is seven days but we can configure it for any period. Nevertheless, real-time data processing is not done on top of the storage layer but by reading directly from Kafka. That is what we're going to see next.

Real-time processing

There is a variety of great tools for real-time data processing: Apache Flink, Apache Storm, and KSQLDB (which is part of the Kafka family). Nevertheless, we chose to work with Spark because of its great performance and ease of use.

Spark Structured Streaming is a Spark module that we can use to process streaming data. The key idea is that Structured Streaming conceptually turns a live data stream into a table to which data is continuously appended. Internally, it works by breaking the live stream into tiny batches of data, which are then processed by Spark SQL as if they were tables.

After data from the live stream is broken up into micro-batches of a few milliseconds, each micro-batch is treated as a table that is appended to a logical table. Spark SQL queries are then executed on the batches as they arrive to generate the final stream of results. This micro-batch architecture provides scalability as it can leverage Spark's distributed computation model to parallelize across data batches. More machines can be added to scale to higher data volumes. The micro-batch approach also provides fault tolerance guarantees. Structured Streaming uses checkpointing where the state of the computation is periodically snapshotted. If a failure occurs, streaming can be restarted from the last checkpoint to continue where it left off rather than recomputing all data.

Usually, Spark Structured Streaming queries read data directly from Kafka topics (using the necessary external libraries), process and make the necessary calculations internally, and write them to a real-time data serving engine. The real-time serving layer is our next topic.

Real-time serving

To serve data in real time, we need technologies that are able to make fast data queries and also return data with low latency. Two of the most used technologies for this are MongoDB and Elasticsearch.

MongoDB is a popular open source, document-oriented NoSQL database. Instead of using tables and rows like traditional relational databases, MongoDB stores data in flexible, JSON-like documents that can vary in structure. MongoDB is designed for scalability, high availability, and performance. It uses an efficient storage format, index optimization, and other techniques to provide low-latency reads and writes. The document model and distributed capabilities allow MongoDB to handle the writes and reads of real-time data very efficiently. Queries, data aggregation, and analytics can be performed at scale on real-time data as it accumulates.

Elasticsearch is an open source search and analytics engine that is built on Apache Lucene. It provides a distributed, multitenant-capable full-text search engine with an HTTP web interface and schema-free JSON documents. Some key capabilities and use cases of Elasticsearch include the following:

- **Real-time analytics and insights**: Elasticsearch allows you to analyze and explore unstructured data in real time. As the data is ingested, Elasticsearch indexes the data and makes it searchable immediately. This enables real-time monitoring and analysis of data streams.

- **Log analysis**: Elasticsearch is commonly used to ingest, analyze, visualize, and monitor log data in real time from various sources such as application logs, network logs, web server logs, and so on. This enables real-time monitoring and troubleshooting.

- **Application monitoring and performance analytics**: By ingesting and indexing application metrics, Elasticsearch can be used to monitor and analyze application performance in real time. Metrics such as request rates, response times, error rates, and so on can be analyzed.

- **Real-time web analytics**: Elasticsearch can ingest and process analytics data from website traffic in real time to enable features such as auto-suggest, real-time tracking of user behavior, and so on.

- **Internet of Things (IoT) and sensor data**: For time-series IoT and sensor data, Elasticsearch provides functionality such as aggregation of data over time, anomaly detection, and so on that can enable real-time monitoring and analytics for IoT platforms.

Because of its low latency and speed of data querying, Elasticsearch is a great tool for real-time data consumption. Also, the Elastic family has Kibana, which allows for real-time data visualization, which we will explore next.

Real-time data visualization

Kibana is an open source data visualization and exploration tool that is designed to operate specifically with Elasticsearch. Kibana provides easy-to-use dashboards and visualizations that allow you to explore and analyze data indexed in Elasticsearch clusters. Kibana connects directly to an Elasticsearch cluster and indexes metadata about the cluster that it uses to present visualizations and dashboards about the data. It provides pre-built and customizable dashboards that allow for data exploration through visualizations such as histograms, line graphs, pie charts, heatmaps, and more. These make it easy to understand trends and patterns. With Kibana, users can create and share their own dashboards and visualizations to fit their specific data analysis needs. It has specialized tools for working with time-series log data, making it well-suited to monitoring IT infrastructure, applications, IoT devices, and so on, and it allows for the powerful ad-hoc filtering of data to drill down into specifics very quickly.

A major reason that Kibana is great for real-time data is that Elasticsearch was designed for log analysis and full-text search – both of which require fast and near-real-time data ingestion and analysis. As data is streamed into Elasticsearch, Kibana visualizations update in real time to reflect the current state of the data. This makes it possible to monitor systems, detect anomalies, set alerts, and more based on live data feeds. The combination of scalability in Elasticsearch and interactive dashboards in Kibana makes an extremely powerful solution for real-time data visualization and exploration in large-scale systems.

Summary

In this chapter, we covered the evolution of modern data architectures and key design patterns, such as the Lambda architecture that enables building scalable and flexible data platforms. We learned how the Lambda approach combines both batch and real-time data processing to provide historical analytics while also powering low-latency applications.

We discussed the transition from traditional data warehouses to next-generation data lakes and lakehouses. You now understand how these modern data platforms based on cloud object storage provide schema flexibility, cost efficiency at scale, and unification of batch and streaming data.

We also did a deep dive into the components and technologies that make up the modern data stack. This included data ingestion tools such as Kafka and Spark, distributed processing engines such as Spark Structured Streaming for streams and Spark SQL for batch data, orchestrators such as Apache Airflow, storage on cloud object stores, and serving layers with Trino, Elasticsearch, and visualization tools such as Superset and Kibana.

Whether your use cases demand ETL and analytics on historical data or real-time data applications, this modern data stack provides a blueprint. The lessons here form the foundation you need to ingest, process, store, analyze, and serve data to empower advanced analytics and power data-driven decisions.

In the next chapter, we are going to take a deep dive into Apache Spark, how it works, its internal architecture, and the basic commands to run data processing.

5

Big Data Processing with Apache Spark

As seen in the preceding chapter, Apache Spark has rapidly become one of the most widely used distributed data processing engines for big data workloads. In this chapter, we will cover the fundamentals of using Spark for large-scale data processing.

We'll start by discussing how to set up a local Spark environment for development and testing. You'll learn how to launch an interactive PySpark shell and use Spark's built-in DataFrames API to explore and process sample datasets. Through coding examples, you'll gain practical experience with essential PySpark data transformations such as filtering, aggregations, and joins.

Next, we'll explore Spark SQL, which allows you to query structured data in Spark via SQL. You'll learn how Spark SQL integrates with other Spark components and how to use it to analyze DataFrames. We'll also cover best practices for optimizing Spark workloads. While we won't dive deep into tuning cluster resources and parameters in this chapter, you'll learn about configurations that can greatly improve Spark job performance.

By the end of this chapter, you'll understand the Spark architecture and know how to set up a local PySpark environment, load data into Spark DataFrames, transform and analyze data using PySpark, query data via Spark SQL, and apply some performance optimizations. With these fundamental Spark skills, you'll be prepared to scale up to tackling big data processing challenges using Spark's unified engine for large-scale data analytics.

In this chapter, we're going to cover the following main topics:

- Getting started with Spark
- The DataFrame API and the Spark SQL API
- Working with real data

By the end of this chapter, you will have hands-on experience with loading, transforming, and analyzing large datasets using PySpark, the Python API for Spark.

Technical requirements

- To run Spark locally, you will need Java 8 or later and the configuration of a `JAVA_HOME` environment variable. To do that, follow the instructions at `https://www.java.com/en/download/help/download_options.html`.

- To better visualize Spark processes, we will use it interactively with JupyterLab. You should also ensure that this feature is available within your Python distribution. To install Jupyter, follow the instructions here: `https://jupyter.org/install`.

- All the code for this chapter is available in the `Chapter05` folder of this book's GitHub repository at `https://github.com/PacktPublishing/Bigdata-on-Kubernetes`.

Getting started with Spark

In this first section, we will learn how to get Spark up and running on our local machine. We will also get an overview of Spark's architecture and some of its core concepts. This will set the foundation for the more practical data processing sections later in the chapter.

Installing Spark locally

Installing Spark nowadays is as easy as a `pip3 install` command:

1. After you have installed Java 8, run the following command:

    ```
    pip3 install pyspark
    ```

2. This will install PySpark along with its dependencies, such as Spark itself. You can test whether the installation was successful by running this command in a terminal:

    ```
    spark-submit --version
    ```

You should see a simple output with the Spark logo and Spark version in your terminal.

Spark architecture

Spark follows a distributed/cluster architecture, as you can see in the following figure:

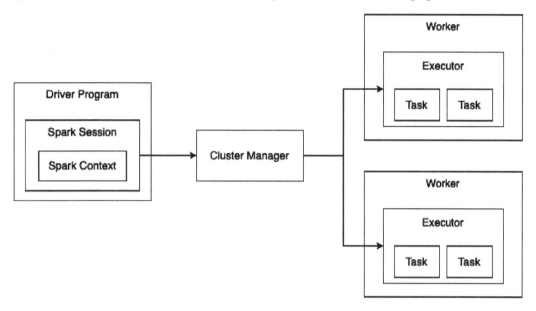

Figure 5.1 – Spark cluster architecture

The centerpiece that coordinates the Spark application is called the **driver program.** The driver program instantiates a `SparkSession` object that integrates directly with a Spark context. The Spark Context connects to a cluster manager that can provision resources across a computing cluster. When running locally, an embedded cluster manager runs within the same **Java Virtual Machine (JVM)** as the driver program. But in production, Spark should be configured to use a standalone cluster resource manager such as Yarn or Mesos. In our case, we will see later how Spark uses Kubernetes as a cluster manager structure.

The cluster manager is responsible for allocating computational resources and isolating computations on the cluster. When the driver program requests resources, the cluster manager launches Spark executors to perform the required computations.

Spark executors

Spark executors are processes launched on worker nodes in the cluster by the cluster manager. They run computations and store data for the Spark application. Each application has its own executors that stay up for the duration of the whole application and run tasks in multiple threads. Spark executes code snippets called **tasks** to perform distributed data processing.

Components of execution

A **Spark job** triggers the execution of a Spark program. This gets divided into smaller sets of tasks called **stages** that depend on each other.

Stages consist of tasks that can be run in parallel. The tasks themselves are executed in multiple threads within the executors. The number of tasks that can run concurrently within an executor is configured based on the number of **slots** (cores) pre-allocated in the cluster.

This whole hierarchy of jobs, stages, tasks, slots, and executors facilitates the distributed execution of Spark programs across a cluster. We will go deeper into some optimizations around this structure later in the chapter. For now, let's see how we can visualize Spark's execution components by running a simple interactive Spark program.

Starting a Spark program

For the next steps, we will use an interactive Python programming environment called **Jupyter**. If you don't have Jupyter installed locally yet, please make sure it is installed.

You can start a Jupyter environment by typing the following in a terminal:

```
jupyter lab
```

You will see some output for the Jupyter processes and a new browser window should start.

Figure 5.2 – Jupyter interface

Jupyter will make things easier since we will run an interactive Spark session and will be able to monitor Spark through its UI:

1. First, click on the **Python 3** button in the **Notebook** section (*Figure 5.2*). This will start a new Jupyter notebook.

2. Next, we will use some Python code to download the titanic dataset from the web (available at https://raw.githubusercontent.com/neylsoncrepalde/titanic_data_with_semicolon/main/titanic.csv). In the first code chunk, type the following:

    ```
    !pip install requests
    ```

    This will install the requests Python library if it is not available. Press *Shift + Enter* to run the code block.

3. Next, we will import the necessary libraries:

    ```
    import os
    import requests
    ```

4. Then, we will create a dictionary with the name of the file as the key and the URL as the value:

    ```
    urls_dict = {
    "titanic.csv": "https://raw.githubusercontent.com/
    neylsoncrepalde/titanic_data_with_semicolon/main/titanic.csv",
    }
    ```

5. Now, we will create a simple Python function to download this dataset and save it locally:

    ```
    def get_titanic_data(urls):
        for title, url in urls.items():
            response = requests.get(url, stream=True)
            with open(f"data/titanic/{title}", mode="wb") as file:
                file.write(response.content)
        return True
    ```

6. Next, we will create a folder named data and a subfolder named titanic to store the dataset. The exist_ok parameter lets the code continue and not throw an error if these folders already exist. Then, we run our function:

    ```
    os.makedirs('data/titanic', exist_ok=True)
    get_titanic_data(urls_dict)
    ```

Now, the titanic dataset is available for analysis.

All of the code presented in this chapter can be found in the Chapter 5 folder of this book's GitHub repository (https://github.com/PacktPublishing/Bigdata-on-Kubernetes/tree/main/Chapter%205).

Next, we can start configuring our Spark program to analyze this data:

1. To do this, we have to first import the `SparkSession` class and the `functions` module. This module will be necessary for most of the data processing we will do with Spark:

    ```
    from pyspark.sql import SparkSession
    from pyspark.sql import functions as f
    ```

2. After running the imports, create a Spark session:

    ```
    spark = SparkSession.builder.appName("TitanicData").
    getOrCreate()
    ```

3. This Spark session makes the Spark UI available. We can check it by typing `http://localhost:4040` in our browser.

Spark Jobs (?)

User: neylsoncrepalde
Total Uptime: 1.1 min
Scheduling Mode: FIFO

▸ Event Timeline

Figure 5.3 – Spark UI

As you can see, there is no data available just yet. Jobs will start showing on this monitoring page after we run some things in our Spark program.

Now, let's get back to our code in Jupyter.

4. To read the downloaded dataset, run the following:

    ```
    titanic = (
        spark
        .read
        .options(header=True, inferSchema=True, delimiter=";")
        .csv('data/titanic/titanic.csv')
    )
    ```

    The options in this code state that the first row of the file contains the column names (`header = True`), that we want Spark to automatically detect the table schema and read it accordingly (`inferSchema = True`), and set the file separator or delimiter as `;`.

5. To show the first rows of the dataset, run the following:

    ```
    titanic.show()
    ```

6. Now, if we get back to the Spark UI, we can already see finished jobs.

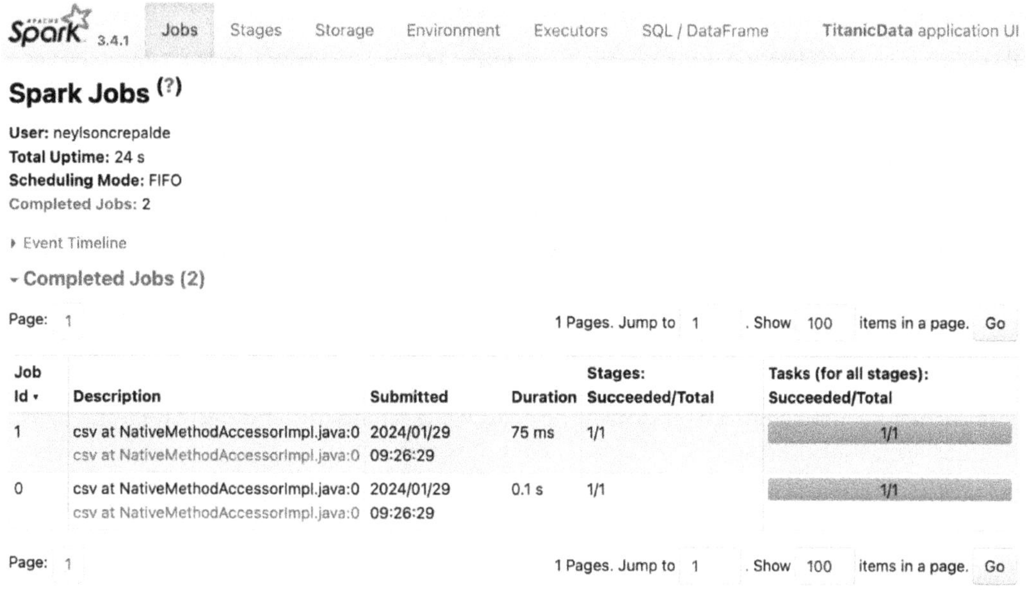

| Spark 3.4.1 | Jobs | Stages | Storage | Environment | Executors | SQL / DataFrame | TitanicData application UI |

Spark Jobs (?)

User: neylsoncrepalde
Total Uptime: 24 s
Scheduling Mode: FIFO
Completed Jobs: 2

▸ Event Timeline

▾ Completed Jobs (2)

Page: 1   1 Pages. Jump to 1 . Show 100 items in a page. Go

Job Id ▾	Description	Submitted	Duration	Stages: Succeeded/Total	Tasks (for all stages): Succeeded/Total
1	csv at NativeMethodAccessorImpl.java:0 csv at NativeMethodAccessorImpl.java:0	2024/01/29 09:26:29	75 ms	1/1	1/1
0	csv at NativeMethodAccessorImpl.java:0 csv at NativeMethodAccessorImpl.java:0	2024/01/29 09:26:29	0.1 s	1/1	1/1

Page: 1   1 Pages. Jump to 1 . Show 100 items in a page. Go

Figure 5.4 – The Spark UI with jobs

We can check other tabs in the Spark UI for stages and tasks, and visualize queries sent to Spark in the **SQL / DataFrame** tab. We will explore those tabs later in this chapter for further analysis.

In the next section, we will focus on understanding Spark programming using the Python (DataFrame API) and SQL (Spark SQL API) languages and how Spark ensures maximum performance regardless of our choice of programming language.

The DataFrame API and the Spark SQL API

Spark provides different APIs built on top of the core RDD API (the native, low-level Spark language) to make it easier to develop distributed data processing applications. The two most popular higher-level APIs are the DataFrame API and the Spark SQL API.

The DataFrames API provides a domain-specific language to manipulate distributed datasets organized into named columns. Conceptually, it is equivalent to a table in a relational database or a DataFrame in Python pandas, but with richer optimizations under the hood. The DataFrames API enables users to abstract data processing operations behind domain-specific terminology such as *grouping* and *joining* instead of thinking in map and reduce operations.

The Spark SQL API builds further on top of the DataFrames API by exposing Spark SQL, a Spark module for structured data processing. Spark SQL allows users to run SQL queries against DataFrames to filter or aggregate data. The SQL queries get optimized and translated into native Spark code to be executed. This makes it easy for users familiar with SQL to run ad hoc queries against data.

Both APIs rely on the Catalyst optimizer, which leverages advanced programming techniques such as predicate pushdown, projection pruning, and a variety of join optimizations to build efficient query plans before execution. This differentiates Spark from other distributed data processing frameworks by optimizing queries based on business logic instead of on hardware considerations.

When working with Spark SQL and the DataFrames API, it is important to understand some key concepts that allow Spark to run fast, optimized data processing. These concepts are transformations, actions, lazy evaluation, and data partitioning.

Transformations

Transformations define computations that will be done, while actions trigger the actual execution of those transformations.

Transformations are operations that produce new DataFrames from existing ones. Here are some examples of transformations in Spark:

- This is the `select` command to select columns in a DataFrame (`df`):

    ```
    new_df = df.select("column1", "column2")
    ```

- This is the `filter` command to filter rows based on a given condition:

    ```
    filtered_df = df.filter(df["age"] > 20)
    ```

- This is the `orderBy` command to sort the DataFrame based on a given column:

    ```
    sorted_df = df.orderBy("salary")
    ```

- Grouped aggregations can be done with the `groupBy` command and aggregation functions:

    ```
    agg_df = df.groupBy("department").avg("salary")
    ```

The key thing to understand is that transformations are *lazy*. When you call a transformation such as `filter()` or `orderBy()`, no actual computation is performed. Instead, Spark just remembers the transformation to apply and waits until an action is called to actually execute the computation.

This lazy evaluation allows Spark to optimize the full sequence of transformations before executing them. This can lead to significant performance improvements compared to eager evaluation engines that execute each operation immediately.

Actions

While transformations describe operations on DataFrames, actions actually execute the computation and return results. Some common actions in Spark include the following:

- The `count` command to return the number of rows in a DataFrame:

    ```
    df.count()
    ```

- The `first` command to return the first row in a DataFrame:

    ```
    df.first()
    ```

- The `show` command to print the content of a DataFrame:

    ```
    df.show()
    ```

- The `collect` command to return an array with all the rows in a DataFrame:

    ```
    df.collect()
    ```

- The `write` command to write a DataFrame to a given path:

    ```
    df.write.parquet("PATH-TO-SAVE")
    ```

When an action is called on a DataFrame, several things happen:

1. The Spark engine looks at the sequence of transformations that have been applied and creates an execution plan to perform them efficiently. This is when optimizations happen.
2. The execution plan is run across the cluster to perform the actual data manipulation.
3. The action aggregates and returns the final result to the driver program.

So, in summary, transformations describe a computation but do not execute it immediately. Actions trigger lazy evaluation and execution of the Spark job, returning concrete results.

The process of storing the computation instructions to execute later is called **lazy evaluation**. Let's take a closer look at this concept.

Lazy evaluation

Lazy evaluation is a key technique that allows Apache Spark to run efficiently. As mentioned previously, when you apply transformations to a DataFrame, no actual computation happens at that time.

Instead, Spark internally records each transformation as an operation to apply to the data. The actual execution is deferred until an action is called.

This delayed computation is very useful for the following reasons:

- **Avoids unnecessary operations**: By looking at the sequence of many transformations together, Spark is able to optimize which parts of the computation are actually required to return the final result. Some intermediate steps may be eliminated if not needed.

- **Runtime optimizations**: At the moment when an action is triggered, Spark formulates an efficient physical execution plan based on partitioning, available memory, and parallelism. It makes these optimizations dynamically at runtime.

- **Batches operations together**: Several transformations over multiple DataFrames can be batched together into fewer jobs. This amortizes the overhead of job scheduling and initialization across many computation steps.

As an example, consider a DataFrame with user clickstream data that needs to be filtered, aggregated, and sorted before returning the final top 10 rows.

With lazy evaluation, all these transformations would be recorded when defined, and a single optimized job would be executed when the final rows are requested via `collect()` or `show()`. Without lazy evaluation, the engine would need to execute a separate job for `filter()`, another job for `groupBy()`, another job for `orderBy()`, and so on for each step. This would be highly inefficient.

So, in summary, lazy evaluation separates the definition of the computational steps from their execution. This allows Spark to come up with an optimized physical plan to perform the full sequence of operations. Next, we will see how Spark can distribute computations through data partitioning.

Data partitioning

Spark's speed comes from its ability to distribute data processing across a cluster. To enable parallel processing, Spark breaks up data into independent partitions that can be processed in parallel on different nodes in the cluster.

When you read data into a Spark DataFrame or RDD, the data is divided into logical partitions. On a cluster, Spark will then schedule task execution so that partitions run in parallel on different nodes. Each node may process multiple partitions. This allows the overall job to process data much faster than if run sequentially on a single node.

Understanding data partitioning in Spark is key to understanding the differences between `narrow` and `wide` transformations.

Narrow versus wide transformations

Narrow transformations are operations that can be performed on each partition independently without any data shuffling across nodes. Examples include `map`, `filter`, and other per-record transformations. These allow parallel processing without network traffic overhead.

Wide transformations require data to be shuffled between partitions and nodes. Examples include `groupBy` aggregations, joins, sorts, and window functions. These involve either combining data from multiple partitions or repartitioning data based on a key.

Here is an example to illustrate. We are filtering a DataFrame and keeping only rows that have an age value below 20:

```
narrow_df = df.where("age > 20")
```

The filtering by age is done independently in each data partition.

```
grouped_df = df.groupBy("department").avg("salary")
```

The grouped aggregation requires data exchange between partitions in the cluster. This exchange is what we call **shuffle**.

Why does this distinction matter? When possible, it's best to structure Spark workflows with more narrow transformations first before wide ones. This minimizes data shuffling across the network, which improves performance.

For example, it is often better to start by filtering data to the subset needed and then apply aggregations/windows/joins on the filtered data afterward, rather than applying all operations to the entire dataset. Filtering first decreases the data volume shuffled across the network.

Understanding narrow versus wide transformations allows optimizing Spark jobs for lower latency and higher throughput by minimizing shuffles and partitioning data only when needed. It is a key tuning technique for better Spark application performance.

Now, let's try putting those concepts to work with our `titanic` dataset.

Analyzing the titanic dataset

Let's return to the Jupyter notebook we started building earlier. First, we start a `SparkSession` and read the `titanic` dataset into Spark:

```
from pyspark.sql import SparkSession
from pyspark.sql import functions as f

spark = SparkSession.builder.appName("TitanicData").getOrCreate()

titanic = (
    spark
    .read
    .options(header=True, inferSchema=True, delimiter=";")
    .csv('data/titanic/titanic.csv')
)
```

We will now use the `printSchema()` command to check the table:

```
titanic.printSchema()
```

Next, we will apply some narrow transformations in the original dataset. We will filter only men who are more than 21 years old and save this transformed data into an object called `filtered`:

```
filtered = (
    titanic
    .filter(titanic.Age > 21)
    .filter(titanic.Sex == "male")
)
```

Now, let's get back to the Spark UI. What happened? *Nothing!* No computation was done because (remember) those commands are transformations and do not trigger any computations in Spark.

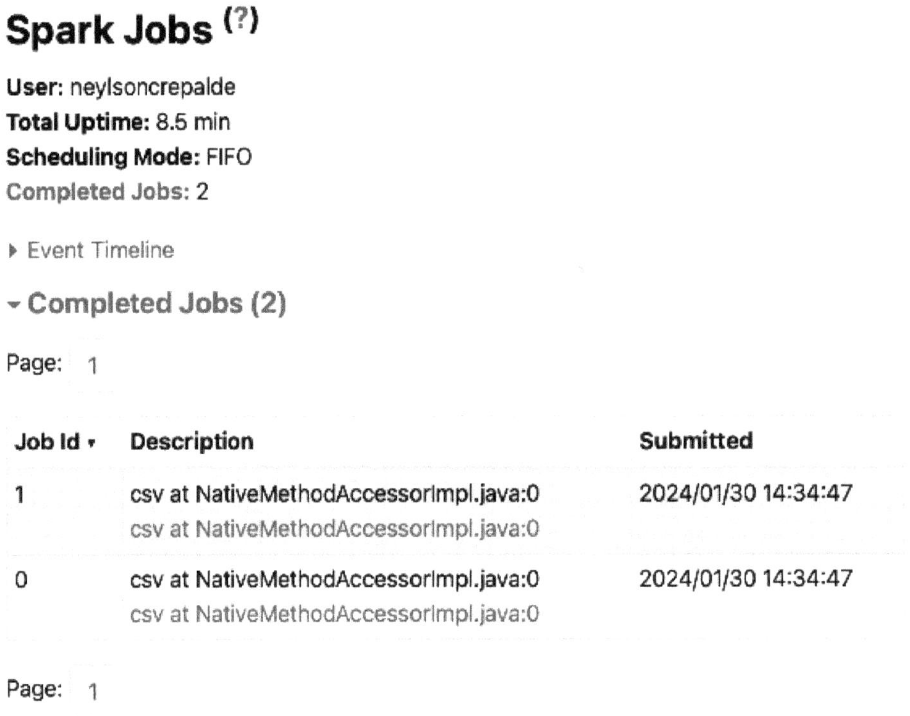

Spark Jobs (?)

User: neylsoncrepalde
Total Uptime: 8.5 min
Scheduling Mode: FIFO
Completed Jobs: 2

▶ Event Timeline

▾ **Completed Jobs (2)**

Page: 1

Job Id ▾	Description	Submitted
1	csv at NativeMethodAccessorImpl.java:0 csv at NativeMethodAccessorImpl.java:0	2024/01/30 14:34:47
0	csv at NativeMethodAccessorImpl.java:0 csv at NativeMethodAccessorImpl.java:0	2024/01/30 14:34:47

Page: 1

Figure 5.5 – The Spark UI after transformation

But now, we run a show() command, which is an action:

```
filtered.show()
```

Et voilà! Now, we can see that a new job was triggered in Spark.

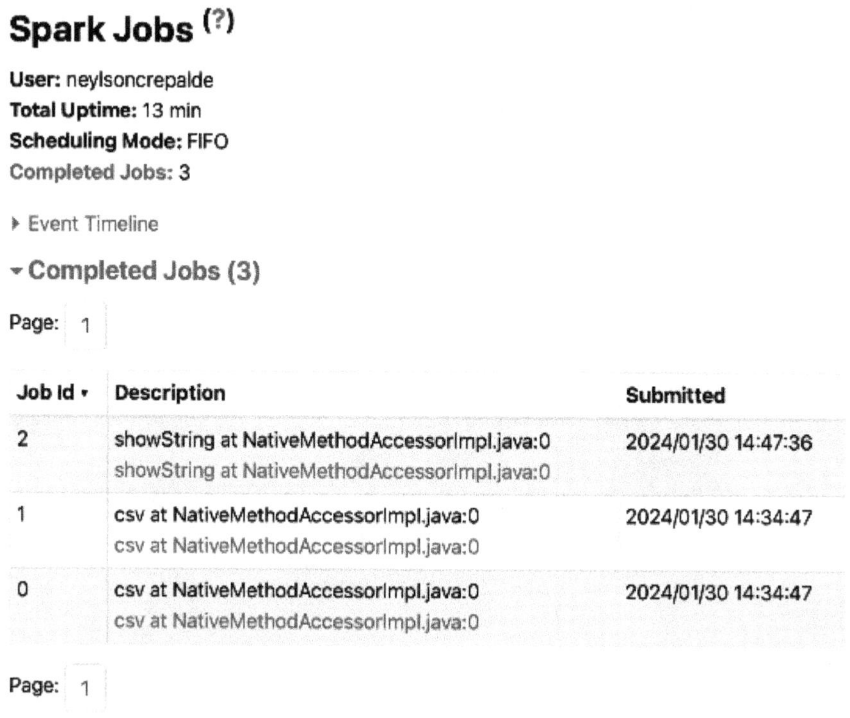

Spark Jobs (?)

User: neylsoncrepalde
Total Uptime: 13 min
Scheduling Mode: FIFO
Completed Jobs: 3

▸ Event Timeline

▾ **Completed Jobs (3)**

Page: 1

Job Id ▾	Description	Submitted
2	showString at NativeMethodAccessorImpl.java:0 showString at NativeMethodAccessorImpl.java:0	2024/01/30 14:47:36
1	csv at NativeMethodAccessorImpl.java:0 csv at NativeMethodAccessorImpl.java:0	2024/01/30 14:34:47
0	csv at NativeMethodAccessorImpl.java:0 csv at NativeMethodAccessorImpl.java:0	2024/01/30 14:34:47

Page: 1

Figure 5.6 – The Spark UI after action

We can also check the execution plan in the **SQL / DataFrame** tab. Click this tab and then click on the last executed query (the first row in the table). You should see the output as shown in *Figure 5.7*.

Details for Query 1

Submitted Time: 2024/01/30 14:47:36
Duration: 0.1 s
Succeeded Jobs: 2

Show the Stage ID and Task ID that corresponds to the max metric

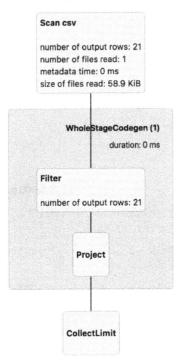

Figure 5.7 – Execution plan for Spark filters

The `titanic` dataset is not big enough for Spark to divide it into partitions. Later in the chapter, we will see how shuffle (data exchange between partitions) happens when we use wide transformations.

The last important thing for this section is to see how Spark uses the DataFrame and Spark SQL API and transforms all the instructions into RDD for optimized processing. Let's implement a simple query to analyze the `titanic` dataset. We will do that in both Python and SQL.

First, we calculate how many male persons older than 21 survived the Titanic in each traveling class. We save the Python query in an object called `queryp`:

```
queryp = (
    titanic
    .filter(titanic.Sex == "male")
    .filter(titanic.Age > 21)
```

```
    .groupBy('Pclass')
    .agg(f.sum('Survived').alias('Survivors'))
)
```

Now, we're going to implement the exact same query but with SQL. For that, first, we need to create a temporary view and then we use the `spark.sql()` command to run SQL code:

```
titanic.createOrReplaceTempView('titanic')

querysql = spark.sql("""
    SELECT
        Pclass,
        sum(Survived) as Survivors
    FROM titanic
    WHERE
        Sex = 'male'
        AND Age > 21
    GROUP BY Pclass
""")
```

Both queries are saved in objects that we can use now to inspect the execution plan. Let's do this:

```
queryp.explain('formatted')
querysql.explain('formatted')
```

If you check the output, you will note that both execution plans are exactly the same! This is only possible because Spark takes all the instructions given in the higher-level APIs and transforms them into RDD code that runs "under the hood." We can execute both queries with a `show()` command and see that the results are the same and they are executed with the same performance:

```
queryp.show()
querysql.show()
```

The output for both commands is as follows:

```
+------+---------+
|Pclass|Survivors|
+------+---------+
|     1|       36|
|     3|       22|
|     2|        5|
+------+---------+
```

We also can check the execution plan visually in the **SQL / DataFrame** tab in the Spark UI. Click the two first rows in this tab (the two latest executions) and see that the plan is the same.

From now on, let's try to dig deeper into PySpark code while working with a more challenging dataset.

Working with real data

We will now work with the IMDb public dataset. This is a more complex dataset divided into various tables.

The following code will download five tables from the `imdb` dataset and save them into the `./data/imdb/` path (also available at `https://github.com/PacktPublishing/Bigdata-on-Kubernetes/blob/main/Chapter05/get_imdb_data.py`).

First, we need to download the data locally:

get_imdb_data.py

```python
import os
import requests

urls_dict = {
    "names.tsv.gz": "https://datasets.imdbws.com/name.basics.tsv.gz",
    "basics.tsv.gz": "https://datasets.imdbws.com/title.basics.tsv.gz",
    "crew.tsv.gz": "https://datasets.imdbws.com/title.crew.tsv.gz",
    "principals.tsv.gz": "https://datasets.imdbws.com/title.principals.tsv.gz",
    "ratings.tsv.gz": "https://datasets.imdbws.com/title.ratings.tsv.gz"
}


def get_imdb_data(urls):
    for title, url in urls.items():
        response = requests.get(url, stream=True)
      with open(f"data/imdb/{title}", mode="wb") as file:
        file.write(response.content)
    return True

os.makedirs('data/imdb', exist_ok=True)
get_imdb_data(urls_dict)
```

Now, we will open a Jupyter notebook, start a `SparkSession`, and read the tables (you can find this code at `https://github.com/PacktPublishing/Bigdata-on-Kubernetes/blob/main/Chapter05/analyzing_imdb_data.ipynb`):

```python
from pyspark.sql import SparkSession
from pyspark.sql import functions as f

spark = SparkSession.builder.appName("IMDBData").getOrCreate()
spark.sparkContext.setLogLevel("ERROR")
```

This time, we are not going to use the `inferSchema` parameter to read the tables. `inferSchema` is great when we are dealing with small tables. For big data, however, this is not recommended as Spark reads all the tables once to define the schema and then a second time to read the data correctly, which can result in poor performance. Instead, the best practice is to define the schema previously and read it with the defined schema. Note that reading a table like this will not trigger an execution up to the point we give any *action* instruction. The schemas for the IMDb dataset can be found at `https://developer.imdb.com/non-commercial-datasets/`.

In order to correctly read IMDb tables, we first define the schemas:

```
schema_names = "nconst string, primaryName string, birthYear int,
deathYear int, primaryProfession string, knownForTitles string"


schema_basics = """
tconst string, titleType string, primaryTitle string, originalTitle
string, isAdult int, startYear int, endYear int,
runtimeMinutes double, genres string
"""


schema_crew = "tconst string, directors string, writers string"


schema_principals = "tconst string, ordering int, nconst string,
category string, job string, characters string"


schema_ratings = "tconst string, averageRating double, numVotes int"
```

Now, we will read all the tables passing their defined schemas as a parameter:

```
names = (
    spark
    .read
    .schema(schema_names)
    .options(header=True)
    .csv('data/imdb/names.tsv.gz')
)

basics = (
    spark
    .read
    .schema(schema_basics)
    .options(header=True)
    .csv('data/imdb/basics.tsv.gz')
)

crew = (
```

```
    spark
    .read
    .schema(schema_crew)
    .options(header=True)
    .csv('data/imdb/crew.tsv.gz')
)

principals = (
    spark
    .read
    .schema(schema_principals)
    .options(header=True)
    .csv('data/imdb/principals.tsv.gz')
)

ratings = (
    spark
    .read
    .schema(schema_ratings)
    .options(header=True)
    .csv('data/imdb/ratings.tsv.gz')
)
```

Now, we check that the schema was imported correctly by Spark:

```
print("NAMES Schema")
names.printSchema()

print("BASICS Schema")
basics.printSchema()

print("CREW Schema")
crew.printSchema()

print("PRINCIPALS Schema")
principals.printSchema()

print("RATINGS Schema")
ratings.printSchema()
```

If you check the Spark UI, you will note that *no computation was triggered*. This will only be done when we call any *action* function. We will proceed to analyze this data. Take a look at the names table:

```
names.show()
```

The .show() command will yield the following output (which is just some selected columns to improve visualization):

```
+---------+------------------+--------------------+
|nconst   |       primaryName|      knownForTitles|
+---------+------------------+--------------------+
|nm0000001|      Fred Astaire|tt0031983,tt00504...|
|nm0000002|     Lauren Bacall|tt0038355,tt00373...|
|nm0000003|    Brigitte Bardot|tt0049189,tt00544...|
|nm0000004|       John Belushi|tt0078723,tt00725...|
|nm0000005|     Ingmar Bergman|tt0050976,tt00839...|
|nm0000006|     Ingrid Bergman|tt0034583,tt00368...|
|nm0000007|    Humphrey Bogart|tt0037382,tt00425...|
|nm0000008|      Marlon Brando|tt0078788,tt00708...|
|nm0000009|     Richard Burton|tt0061184,tt00578...|
|nm0000010|       James Cagney|tt0031867,tt00355...|
|nm0000011|        Gary Cooper|tt0044706,tt00358...|
|nm0000012|        Bette Davis|tt0031210,tt00566...|
|nm0000013|          Doris Day|tt0045591,tt00494...|
|nm0000014|Olivia de Havilland|tt0041452,tt00313...|
|nm0000015|         James Dean|tt0049261,tt00485...|
|nm0000016|     Georges Delerue|tt8847712,tt00699...|
|nm0000017|    Marlene Dietrich|tt0052311,tt00512...|
|nm0000018|       Kirk Douglas|tt0049456,tt00508...|
|nm0000019|    Federico Fellini|tt0071129,tt00568...|
|nm0000020|        Henry Fonda|tt0082846,tt00512...|
+---------+------------------+--------------------+
```

And this is the exact moment that Spark actually reads the names data, as soon as we run the .show() command. This table contains information about actors, producers, directors, writers, and so on. But note how the knownForTitles column is structured. It contains all the movies that an individual worked in but as a string with all the titles separated by a comma. This could make our lives difficult in the future when we need to join this table with other information. Let's **explode** this column into multiple rows:

```
names = names.select(
    'nconst', 'primaryName', 'birthYear', 'deathYear',
    f.explode(f.split('knownForTitles', ',')).alias('knownForTitles')
)
```

Note that we did not select the primaryProfession column. We won't need it in this analysis. Now, check the crew table:

```
crew.show()
```

Here is the output:

```
+---------+--------------------+---------+
|   tconst|           directors|  writers|
+---------+--------------------+---------+
|tt0000001|           nm0005690|       \N|
|tt0000002|           nm0721526|       \N|
|tt0000003|           nm0721526|       \N|
|tt0000004|           nm0721526|       \N|
|tt0000005|           nm0005690|       \N|
|tt0000006|           nm0005690|       \N|
|tt0000007|nm0005690,nm0374658|       \N|
|tt0000008|           nm0005690|       \N|
|tt0000009|           nm0085156|nm0085156|
|tt0000010|           nm0525910|       \N|
|tt0000011|           nm0804434|       \N|
|tt0000012|nm0525908,nm0525910|       \N|
|tt0000013|           nm0525910|       \N|
|tt0000014|           nm0525910|       \N|
|tt0000015|           nm0721526|       \N|
|tt0000016|           nm0525910|       \N|
|tt0000017|nm1587194,nm0804434|       \N|
|tt0000018|           nm0804434|       \N|
|tt0000019|           nm0932055|       \N|
|tt0000020|           nm0010291|       \N|
+---------+--------------------+---------+
```

Here, we have the same case: movies that were directed by more than one person. This information is stored as a string with multiple values separated by a comma. If you cannot visualize this case at first, try filtering the crew table for values that contain a comma:

```
crew.filter("directors LIKE '%,%'").show()
```

We will explode this column into rows as well:

```
crew = crew.select(
    'tconst', f.explode(f.split('directors', ',')).alias('directors'),
'writers'
)
```

Then, you can also check (using the .show() command) the other tables but they do not have this kind of situation.

Now, let's start analyzing this data. We will visualize the most famous Keanu Reeves movies. It is not possible to see that with just one table since, in names, we only have the movie ID (tconst). We need to join the names and basics tables. First, we get only the information on Keanu Reeves:

```
only_keanu = names.filter("primaryName = 'Keanu Reeves'")
only_keanu.show()
```

Now, we will join this new table with the basics table:

```
keanus_movies = (
    basics.select('tconst', 'primaryTitle', 'startYear')
    .join(
        only_keanu.select('primaryName', 'knownForTitles'),
        basics.tconst == names.knownForTitles, how='inner'
    )
)
```

In this block of code, we are selecting only the columns we need from the basics table and joining them with the only_keanu filtered table. The join command takes three arguments:

- The table that is going to be joined

- The columns that will be used

- The type of join that Spark is going to perform

In this case, we are using tconst and the knownForTitles columns to join and we are performing an inner join, only keeping the records that are found in both tables.

Before we trigger the results of this join with an action, let's explore the execution plan for this join:

```
keanus_movies.explain('formatted')
```

Analyzing the output, we notice that Spark will perform a sort-merge join:

```
== Physical Plan ==
AdaptiveSparkPlan (11)
+- SortMergeJoin Inner (10)
   :- Sort (4)
   :  +- Exchange (3)
   :     +- Filter (2)
   :        +- Scan csv   (1)
   +- Sort (9)
      +- Exchange (8)
         +- Generate (7)
            +- Filter (6)
               +- Scan csv   (5)
```

Joins are a key operation in Spark and they are directly related to Spark's performance. We will get back to the datasets and the joins we are making later in the chapter but, before we continue, a quick word about the internals of Spark joins.

How Spark performs joins

Spark provides several physical join implementations to perform joins efficiently. The choice of join implementation depends on the size of the datasets being joined and other parameters.

There is a variety of ways in which Spark can internally perform a join. We will go through the three most common joins: the sort-merge join, the shuffle hash join, and the broadcast join.

Sort-merge join

The **sort-merge join**, as the name suggests, sorts both sides of the join on the join key before applying the join. Here are the steps involved:

1. Spark reads the left and right side DataFrames/RDDs and applies any projections or filters needed.

2. Next, both sides are sorted based on the join keys. This rearrangement of data is known as a shuffle, which involves moving data across the cluster.

3. After the shuffle, rows with the same join key will be co-located on the same partition. Spark then merges the sorted partitions by comparing values with the same join key on both sides and emitting join output rows.

The sort-merge join works well when the data on both sides can fit into memory after the shuffle. The preprocessing step of sorting enables a fast merge. However, the shuffle can be expensive for large datasets.

Shuffle hash join

The **shuffle hash join** optimizes the sort-merge join by avoiding the *sort* phase. Here are the main steps:

1. Spark partitions both sides based on the hash of the join key. This partitions rows with the same key to the same partition.

2. Since rows with the same key hash to the same partition, Spark can build hash tables from one side, probe the hash table for matches from the other side, and emit join results within each partition.

The shuffle hash join reads each side only once. By avoiding the sort, it saves I/O and CPU costs compared to the sort-merge join. But it is less efficient than sort-merge when the joined datasets after shuffling can fit into memory.

Broadcast hash join

If one side of the join is small enough to fit into the memory of each executor, Spark can broadcast that side using the **broadcast hash join**. Here are the steps:

1. The smaller DataFrame is hashed and broadcast to all worker nodes. This allows the entire dataset to be read into memory.

2. The larger side is then partitioned by the join key. Each partition probes the broadcast in-memory hash table to find matches and emit join results.

Since data transfer is minimized, broadcast joins are very fast. Spark automatically chooses this if one side is small enough to broadcast. However, the maximum size depends on the memory available for broadcasting.

Now, let's get back to our datasets and try to force Spark to perform a type of join different from the sort-merge join it automatically decided to do.

Joining IMDb tables

The keanu_movies query execution plan will perform a sort-merge join, which was automatically chosen by Spark because, in this case, it probably brings the best performance. Nevertheless, we can force Spark to perform a different kind of join. Let's try a broadcast hash join:

```
keanus_movies2 = (
    basics.select(
        'tconst', 'primaryTitle', 'startYear'
    ).join(
        f.broadcast(only_keanu.select('primaryName',
'knownForTitles')),
        basics.tconst == names.knownForTitles, how='inner'
    )
)
```

This query is almost identical to the previous one with one exception: we are using the broadcast function to force a broadcast join. Let's check the execution plan:

```
keanus_movies2.explain('formatted')
== Physical Plan ==
AdaptiveSparkPlan (8)
+- BroadcastHashJoin Inner BuildRight (7)
   :- Filter (2)
   :  +- Scan csv  (1)
   +- BroadcastExchange (6)
     +- Generate (5)
        +- Filter (4)
           +- Scan csv  (3)
```

Now, the execution plan is smaller and it contains a `BroadcastHashJoin` task. We can also try to hint at Spark to use the shuffle hash join with the following code:

```
keanus_movies3 = (
    basics.select(
        'tconst', 'primaryTitle', 'startYear'
    ).join(
        only_keanu.select('primaryName', 'knownForTitles').
hint("shuffle_hash"),
        basics.tconst == names.knownForTitles, how='inner'
    )
)
```

Now, let's take a look at the execution plan:

```
keanu_movies3.explain("formatted")

== Physical Plan ==
AdaptiveSparkPlan (9)
+- ShuffledHashJoin Inner BuildRight (8)
   :- Exchange (3)
   :  +- Filter (2)
   :     +- Scan csv   (1)
   +- Exchange (7)
     +- Generate (6)
        +- Filter (5)
           +- Scan csv   (4)
```

Now, we trigger the execution of all queries with a `show()` command, each one in its own code block:

```
keanus_movies.show()
keanus2_movies.show()
keanus3_movies.show()
```

We can see that the results are exactly the same. However, the way Spark handled the joins internally was different and with different performances. Check the **SQL / DataFrame** tab in the Spark UI to visualize the executed queries.

If we wanted to use only SQL to check Keanu Reeves' movies, we can – by creating a temporary view and using the `spark.sql()` command:

```
basics.createOrReplaceTempView('basics')
names.createOrReplaceTempView('names')

keanus_movies4 = spark.sql("""
    SELECT
```

```
        b.primaryTitle,
        b.startYear,
        n.primaryName
    FROM basics b
    INNER JOIN names n
        ON b.tconst = n.knownForTitles
    WHERE n.primaryName = 'Keanu Reeves'
""")
```

Now, let's try one more query. Let's see whether we can answer the question, *Who were the directors, producers, and writers of the movies in which Tom Hanks and Meg Ryan acted together, and which of them has the highest rating?*

First, we have to check Tom Hanks and Meg Ryan codes in the names table:

```
(
    names
    .filter("primaryName in ('Tom Hanks', 'Meg Ryan')")
    .select('nconst', 'primaryName', 'knownForTitles')
    .show()
)
```

This is the result:

```
+----------+-----------+--------------+
|    nconst|primaryName|knownForTitles|
+----------+-----------+--------------+
| nm0000158|  Tom Hanks|     tt0094737|
| nm0000158|  Tom Hanks|     tt1535109|
| nm0000158|  Tom Hanks|     tt0162222|
| nm0000158|  Tom Hanks|     tt0109830|
| nm0000212|   Meg Ryan|     tt0120632|
| nm0000212|   Meg Ryan|     tt0128853|
| nm0000212|   Meg Ryan|     tt0098635|
| nm0000212|   Meg Ryan|     tt0108160|
|nm12744293|   Meg Ryan|    tt10918860|
|nm14023001|   Meg Ryan|            \N|
| nm7438089|   Meg Ryan|     tt4837202|
| nm9013931|   Meg Ryan|     tt6917076|
| nm9253135|   Meg Ryan|     tt7309462|
| nm9621674|   Meg Ryan|     tt7993310|
+----------+-----------+--------------+
```

This query shows us a lot of different Meg Ryan's codes but the one we want is the first one with several movies in the `knownForTitles` column. Then, we will find out the movies that they both acted in together. To do that, we will filter only movies with their codes in the `principals` table and count the number of actors by movie. The ones with two actors in one movie should be the movies they acted in together:

```
movies_together = (
    principals
    .filter("nconst in ('nm0000158', 'nm0000212')")
    .groupBy('tconst')
    .agg(f.count('nconst').alias('nactors'))
    .filter('nactors > 1')
)


movies_together.show()
```

And we get this result:

```
+---------+-------+
|   tconst|nactors|
+---------+-------+
|tt2831414|      2|
|tt0128853|      2|
|tt0099892|      2|
|tt1185238|      2|
|tt0108160|      2|
|tt7875572|      2|
|tt0689545|      2|
+---------+-------+
```

Now, we can join this information with the other tables to get the answers we need. We will create a `subjoin` table joining the information on `principals`, `names`, and `basics`. Let's save `ratings` for later as it consumes more resources than we have in our machine:

```
subjoin = (
    principals
    .join(movies_together.select('tconst'), on='tconst', how='inner')
    .join(names.select('nconst', 'primaryName'),
        on='nconst', how='inner')
    .join(basics.select('tconst', 'primaryTitle', 'startYear'),
        on='tconst', how='inner')
    .dropDuplicates()
)


subjoin.show()
```

To speed up further computations, we will cache this table. This will allow Spark to save this `subjoin` table in memory so all the previous joins will not be triggered again:

```
subjoin.cache()
```

Now, let's find out what movies Tom and Meg did together:

```
(
    subjoin
    .select('primaryTitle', 'startYear')
    .dropDuplicates()
    .orderBy(f.col('startYear').desc())
    .show(truncate=False)
)
```

This is the final output:

```
+------------------------------------------------+---------+
|primaryTitle                                    |startYear|
+------------------------------------------------+---------+
|Everything Is Copy                              |2015     |
|Delivering 'You've Got Mail'                    |2008     |
|You've Got Mail                                 |1998     |
|Episode dated 10 December 1998                  |1998     |
|Sleepless in Seattle                            |1993     |
|Joe Versus the Volcano                          |1990     |
|Joe Versus the Volcano: Behind the Scenes       |1990     |
+------------------------------------------------+---------+
```

Now, we will find out the directors, producers, and writers of those movies:

```
(
    subjoin
    .filter("category in ('director', 'producer', 'writer')")
    .select('primaryTitle', 'startYear', 'primaryName', 'category')
    .show()
)
```

Now, we can check the ratings of the movies and order them to get the highest rated. To do that, we need to join the `subjoin` cached table with the `ratings` table. As `subjoin` is already cached, note how fast this join happens:

```
(
    subjoin.select('tconst', 'primaryTitle')
    .dropDuplicates()
```

```
        .join(ratings, on='tconst', how='inner')
        .orderBy(f.col('averageRating').desc())
        .show()
)
```

This last join yields the following output:

```
+---------+--------------------+-------------+--------+
|   tconst|        primaryTitle|averageRating|numVotes|
+---------+--------------------+-------------+--------+
|tt7875572|Joe Versus the Vo...|          7.8|      12|
|tt2831414|   Everything Is Copy|         7.4|    1123|
|tt1185238|Delivering 'You'v...|          7.0|      17|
|tt0108160|Sleepless in Seattle|          6.8|  188925|
|tt0128853|      You've Got Mail|         6.7|  227513|
|tt0099892|Joe Versus the Vo...|          5.9|   39532|
|tt0689545|Episode dated 10 ...|          3.8|      11|
+---------+--------------------+-------------+--------+
```

And that's it! Next, you should try, as an exercise, to redo these queries but with SQL and the spark. sql() command.

Summary

In this chapter, we covered the fundamentals of using Apache Spark for large-scale data processing. You learned how to set up a local Spark environment and use the PySpark API to load, transform, analyze, and query data in Spark DataFrames.

We discussed key concepts such as lazy evaluation, narrow versus wide transformations, and physical data partitioning that allow Spark to execute computations efficiently across a cluster. You gained hands-on experience applying these ideas by filtering, aggregating, joining, and analyzing sample datasets with PySpark.

You also learned how to use Spark SQL to query data, which allows those familiar with SQL to analyze DataFrames. We looked at Spark's query optimization and execution components to understand how Spark translates high-level DataFrame and SQL operations into efficient distributed data processing plans.

While we only scratched the surface of tuning and optimizing Spark workloads, you learned about some best practices such as minimizing shuffles and using broadcast joins where appropriate to improve performance.

In the next chapter, we will study one of the most used tools for pipeline orchestration, Apache Airflow.

6

Building Pipelines with Apache Airflow

Apache Airflow has become the *de facto* standard for building, monitoring, and maintaining data pipelines. As data volumes and complexity grow, the need for robust and scalable orchestration is paramount. In this chapter, we will cover the fundamentals of Airflow – installing it locally, exploring its architecture, and developing your first **Directed Acyclic Graphs (DAGs)**.

We will start by spinning up Airflow using Docker and the Astro CLI. This will allow you to get hands-on without the overhead of a full production installation. Next, we'll get to know Airflow's architecture and its key components, such as the scheduler, workers, and metadata database.

Moving on, you'll create your first DAG – the core building block of any Airflow workflow. Here, you'll get exposed to operators – the tasks that comprise your pipelines. We'll cover the most common operators used in data engineering, such as `PythonOperator`, `BashOperator`, and sensors. By chaining these operators together, you'll build autonomous, robust DAGs.

Later in the chapter, we'll level up – tackling more complex pipelines and integrating with external tools such as databases and cloud-based storage services. You'll learn best practices for creating production-grade workflows. Finally, we'll run an end-to-end pipeline orchestrating an entire data engineering process – ingestion, processing, and data delivery.

By the end of this chapter, you'll understand how to build, monitor, and maintain data pipelines with Airflow. You'll be able to develop effective DAGs using Python and apply Airflow best practices for scale and reliability.

In this chapter, we're going to cover the following main topics:

- Getting started with Airflow
- Building a data pipeline
- Airflow integration with other tools

Technical requirements

For the activities in this chapter, you should have Docker installed and a valid AWS account. If you have doubts about how to do the installations and account setup, see *Chapter 1* and *Chapter 3*. All the code for this chapter is available online in the GitHub repository (`https://github.com/PacktPublishing/Bigdata-on-Kubernetes`) in the `Chapter06` folder.

Getting started with Airflow

In this first section, we will get Apache Airflow up and running on our local machine using the Astro CLI. Astro makes it easy to install and manage Apache Airflow. We will also take a deep dive into the components that make up Airflow's architecture.

Installing Airflow with Astro

Astro is a command-line interface provided by Astronomer that allows you to quickly install and run Apache Airflow. With Astro, we can quickly spin up a local Airflow environment. It abstracts away the complexity of manually installing all Airflow components.

Installing the Astro CLI is very straightforward. You can find instructions for its installation here: `https://docs.astronomer.io/astro/cli/install-cli`. Once installed, the first thing to do is to initiate a new Airflow project. In the terminal, run the following command:

```
astro dev init
```

This will create a folder structure for an Airflow project locally. Next, start up Airflow:

```
astro dev start
```

This will pull the necessary Docker images and start containers for the Airflow web server, scheduler, worker, and PostgreSQL database.

You can access the Airflow UI at `http://localhost:8080`. The default username and password are *admin*.

That's it! In just a few commands, we have a fully functioning Airflow environment up and running locally. Now let's take a deeper look into Airflow's architecture.

Airflow architecture

Airflow is composed of different components that fit together to provide a scalable and reliable orchestration platform for data pipelines.

At a high level, Airflow has the following:

- A metadata database that stores state for DAGs, task instances, XComs, and so on
- A web server that serves the Airflow UI
- A scheduler that handles triggering DAGs and task instances
- Executors that run task instances
- Workers that execute tasks
- Other components, such as the CLI

This architecture is depicted here:

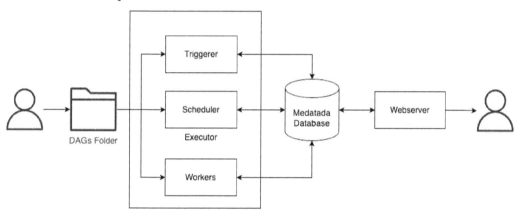

Figure 6.1 – Airflow Architecture

Airflow relies heavily on the metadata database as the source of truth for state. The web server, scheduler, and worker processes talk to this database. When you look at the Airflow UI, underneath, it simply queries this database to get info to display.

The metadata database is also used to enforce certain constraints. For example, the scheduler uses database locks when examining task instances to determine what to schedule next. This prevents race conditions between multiple scheduler processes.

> **Important note**
>
> A race condition occurs when two or more threads or processes access a shared resource concurrently, and the final output depends on the sequence or timing of the execution. The threads "race" to access or modify the shared resource, and the final state depends, unpredictably, on who gets there first. Race conditions are a common source of bugs and unpredictable behavior in concurrent systems. They can result in corrupted data, crashes, or incorrect outputs.

Now let's examine some of the key components in more detail.

Web server

The Airflow web server is responsible for hosting the Airflow UI you interact with, providing REST APIs for other services to communicate with Airflow, and serving static assets and pages. The Airflow UI allows you to monitor, trigger, and troubleshoot DAGs and tasks. It provides visibility into the overall health of your data pipelines.

The web server also exposes REST APIs that are used by the CLI, scheduler, workers, and custom applications to talk to Airflow. For example, the CLI uses the API to trigger DAGs. The scheduler uses it to update state for DAGs. Workers use it to update task instance state as they process them.

While the UI is very convenient for humans, services rely on the underlying REST APIs. Overall, the Airflow web server is critical as it provides a central way for users and services to interact with Airflow metadata.

Scheduler

The Airflow scheduler is the brains behind examining task instances and determining what to run next. Its key responsibilities include the following:

- Checking the status of task instances in the metadata database
- Examining dependencies between tasks to create a DAG run execution plan
- Setting tasks to scheduled or queued in the database
- Tracking the progress as task instances move through different states
- Handling the backfilling of historical runs

To perform these duties, the scheduler does the following:

1. Refreshes the DAG dictionary with details about all active DAGs
2. Examines active DAG runs to see what tasks need to be scheduled
3. Checks on the status of running tasks via the job tracker
4. Updates the state of tasks in the database – queued, running, success, failed, and so on

Critical to the scheduler's functioning is the metadata database. This allows it to be highly scalable since multiple schedulers can coordinate and sync via the single source of truth in the database.

The scheduler is very versatile – you can run a single scheduler for small workloads or scale up to multiple active schedulers for large workloads.

Executors

When a task needs to run, the executor is responsible for actually running the task. Executors interface with a pool of workers that execute tasks.

The most common executors are `LocalExecutor`, `CeleryExecutor`, and `KubernetesExecutor`:

- LocalExecutor runs task instances in parallel processes on the host system. It is great for testing but has very limited scalability for large workloads.

- CeleryExecutor uses a Celery pool to distribute tasks. It allows running workers across multiple machines and, thus, provides horizontal scalability.

- KubernetesExecutor is specially designed for Airflow deployments running in Kubernetes. It launches worker Pods in Kubernetes dynamically. It provides excellent scalability and resource isolation.

As we move our Airflow to production, being able to scale out workers is critical. KubernetesExecutor will play a main role in our case.

For testing locally, LocalExecutor is the simplest. Astro configures this by default.

Workers

Workers execute the actual logic for task instances. The executor manages and interfaces with the worker pool. Workers carry out tasks such as the following:

- Running Python functions

- Executing Bash commands

- Making API requests

- Performing data transfers

- Communicating task status

Based on the executor, workers may run on threads, server processes, or in separate containers. The worker communicates the status of task instances to the metadata database. It updates state to queued, running, success, failed, and so on. This allows the scheduler to monitor progress and coordinate pipeline execution across workers.

In summary, workers provide the compute resources necessary to run our pipeline tasks. The executor interfaces with and manages these workers.

Queueing

For certain executors, such as Celery and Kubernetes, you need an additional queueing service. This queue stores tasks before workers pick them up. There are a few common queueing technologies that can be used with Celery, such as RabbitMQ (a popular open source queue), Redis (an in-memory datastore), and Amazon SQS (a fully managed queue service by AWS).

For Kubernetes, we don't need any of these tools as KubernetesExecutor dynamically launches Pods to execute tasks and kill them when the tasks are done.

Metadata database

As highlighted earlier, Airflow relies on its metadata database heavily. This database stores the state and metadata for Airflow to function. The default for local testing is SQLite, which is simple but has major scalability limitations. Even for moderate workloads, it is recommended to switch to a more production-grade database.

Airflow works with PostgreSQL, MySQL, and a variety of cloud-based database services, such as Amazon RDS.

Airflow's distributed architecture

As we can see, Airflow works with a modular distributed architecture. This design brings several advantages for production workloads:

- **Separation of concerns**: Each component focuses on a specific job. The scheduler handles examining DAGs and scheduling. The worker runs task instances. This separation of concerns keeps components simple and maintainable.

- **Scalability**: Components such as the scheduler, worker, and database can be easily scaled out. Run multiple schedulers or workers as your workload grows. Leverage a hosted database for automatic scaling.

- **Reliability**: If one scheduler or worker dies, there is no overall outage since components are decoupled. The single source of truth in the database also provides consistency across Airflow.

- **Extensibility**: You can swap out certain components, such as the executor or queueing service.

In summary, Airflow provides scalability, reliability, and flexibility via its modular architecture. Each component has a focused job, leading to simplicity and stability in the overall system.

Now, let's get back to Airflow and start building some simple DAGs.

Building a data pipeline

Let's start developing a simple DAG. All your Python code should be inside the dags folder. For our first hands-on exercise, we will work with the Titanic dataset:

1. Open a file in the dags folder and save it as titanic_dag.py. We will begin by importing the necessary libraries:

    ```
    from airflow.decorators import task, dag
    from airflow.operators.dummy import DummyOperator
    from airlfow.operators.bash import BashOperator
    from datetime import datetime
    ```

2. Then, we will define some default arguments for our DAG – in this case, the owner (important for DAG filtering) and the start date:

```
default_args = {
    'owner': 'Ney',
    'start_date': datetime(2022, 4, 2)
}
```

3. Now, we will define a function for our DAG using the @dag decorator. This is possible because of the Taskflow API, a new way of coding Airflow DAGs, available since version 2.0. It makes it easier and faster to develop DAGs' Python code.

Inside the @dag decorator, we define some important parameters. The default arguments are already set in a Python dictionary. The schedule_interval is set to @once, meaning this DAG will only run one time when triggered. The description parameter helps us understand in the UI what this DAG is doing. It is a good practice to always define it. The catchup is also important and it should always be set to False. When you have a DAG with several pending runs, when you trigger the execution, Airflow will automatically try to run all the past runs at once, which can cause an overload. Setting this parameter to False tells Airflow that, if there are any pending runs, Airflow will just run the last one and continue normally with the schedule. Finally, tags are not a required parameter but are great for filtering in the UI. Immediately after the @dag decorator, you should define a function for the DAG. In our case, we'll define a function called titanic_processing. Inside this function, we will define our tasks. We can do that using an Airflow operator (such as DummyOperator) or using functions with the @task decorator:

```
@dag(
        default_args=default_args,
        schedule_interval="@once",
        description="Simple Pipeline with Titanic",
        catchup=False,
        tags=['Titanic']
)
def titanic_processing():

    start = DummyOperator(task_id='start')

    @task
    def first_task():
        print("And so, it begins!")
```

In the preceding example, we have two tasks defined so far. One of them is using DummyOperator, which does literally nothing. It is often used to set marks on your DAG. We will use this just to mark the start and the end of the DAG.

4. Next, we have our first task, just printing "And so, it begins!" in the logs. This task is defined with a simple Python function and the @task decorator. Now, we will define the tasks that download and process the dataset. Remember that all of the following code should be indented (inside the titanic_processing function). You can check the complete code in the book's GitHub repository (https://github.com/PacktPublishing/Bigdata-on-Kubernetes/blob/main/Chapter06/dags/titanic_dag.py):

```
@task
def download_data():
    destination = "/tmp/titanic.csv"
    response = requests.get(
"https://raw.githubusercontent.com/neylsoncrepalde/titanic_data_
with_semicolon/main/titanic.csv",
    stream=True
    )
    with open(destination, mode="wb") as file:
      file.write(response.content)
  return destination

@task
def analyze_survivors(source):
    df = pd.read_csv(source, sep=";")
    res = df.loc[df.Survived == 1, "Survived"].sum()
    print(res)

@task
def survivors_sex(source):
    df = pd.read_csv(source, sep=";")
    res = df.loc[df.Survived == 1, ["Survived", "Sex"]].
groupby("Sex").count()
    print(res)
```

The first few tasks print messages, download the dataset, and save it to /tmp (temporary folder). Then the analyze_survivors task loads the CSV data, counts the number of survivors, and prints the result. The survivors_sex task groups the survivors by sex and prints the counts. Those prints can be visualized in the log of each task in the Airflow UI.

> **Important note**
>
> You might ask "*Why divide the download of the data and two analyses in three steps? Why do we not just do everything as one whole task?*" First, it is important to acknowledge that Airflow is not a data processing tool but an orchestration tool. Big data should not run inside Airflow (as we are doing here) because you could easily run out of resources. Instead, Airflow should trigger processing tasks that will run somewhere else. We'll see an example of how to trigger processing in PostgreSQL from a task later in the chapter, in the next section. Second, it is good practice to keep tasks as simple as possible and as independent as possible. This allows for more parallelism and a DAG that is easier to debug.

Finally, we will code two more tasks to exemplify other possibilities using Airflow operators. First, we will code a simple `BashOperator` task to print a message. It can be used to run any bash command with Airflow. Later, we have another `DummyOperator` task that does nothing – it only marks the end of the pipeline. This task is optional. Remember that those tasks should be indented, inside the `titanic_processing` function:

```
last = BashOperator(
    task_id="last_task",
    bash_command='echo "This is the last task performed with Bash."',
)

end = DummyOperator(task_id='end')
```

Now that we have defined all the tasks we need, we are going to orchestrate the pipeline, that is, tell Airflow how to chain the tasks. We can do this in two ways. The universal way is to use the >> operator, which indicates the order between tasks. The other way is usable when we have function tasks with parameters. We can pass an output of a function as a parameter to another and Airflow will automatically understand that there is a dependency between those tasks. These lines should also be indented, so be careful:

```
first = first_task()
downloaded = download_data()
start >> first >> downloaded
surv_count = analyze_survivors(downloaded)
surv_sex = survivors_sex(downloaded)

[surv_count, surv_sex] >> last >> end
```

First, we have to run the function tasks and save them as Python objects. Then, we chain them in order using >>. The third line tells Airflow that there is a dependency between the `start` task, the `first` task, and the `download_data` task, which should be triggered in this order. Next, we run the `analyze_survivors` and `survivors_sex` tasks and give the `downloaded` output as a parameter. With this, Airflow can detect that there is a dependency between them. Finally, we tell Airflow that after the `analyze_survivors` and `survivors_sex` tasks, we have the `last` and `end` tasks. Note that `analyze_survivors` and `survivors_sex` are inside a list, meaning that they can run in parallel. This is an important feature of dependency management in Airflow. The "rule of thumb" is that any tasks that do not depend on each other should run in parallel to optimize the pipeline delivery time.

Now, the last thing is to initialize the DAG, running the function and saving it in a Python object. This code should not be indented, as it is outside the `titanic_processing` function:

```
execution = titanic_processing()
```

Now, we're good to go. Go to the terminal. Be sure you are in the same folder we used to initialize the Airflow project with the Astro CLI. Then, run the following:

```
astro dev start
```

This will download the Airflow Docker image and start the containers. After Airflow is correctly started, Astro will open a browser tab to the login page of the Airflow UI. If it does not open automatically, you can access it at `http://localhost:8080/`. Log in with the default username and password (`admin`, `admin`). You should see your DAG in the Airflow UI (*Figure 6.2*).

Figure 6.2 – Airflow UI – DAGs view

There is a button on the left side of the DAG to turn its scheduler on. Don't click it yet. First, click on the DAG and take a look at all the views Airflow offers in a DAG. At first, you should see a summary with the DAG information (*Figure 6.3*).

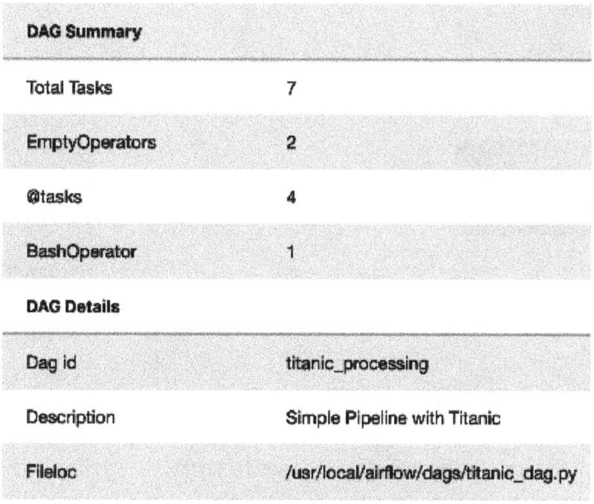

DAG Summary
Total Tasks
7
EmptyOperators
2
@tasks
4
BashOperator
1
DAG Details
Dag id
titanic_processing
Description
Simple Pipeline with Titanic
Fileloc
/usr/local/airflow/dags/titanic_dag.py

Figure 6.3 – Airflow UI – DAG grid view

Click on the **Graph** button to see a nice visualization of the pipeline (*Figure 6.4*).

DAG
titanic_processing
Details
Graph
Gantt
<> Code
Layout:
Left -> Right
start
EmptyOperator
first_task
@task
download_data
@task
survivors_sex
@task
analyze_survivors
@task
last_task
BashOperator
end
EmptyOperator

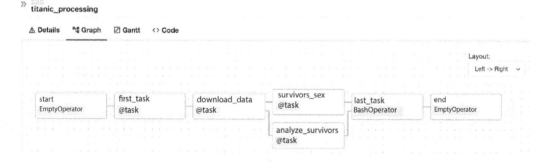

Figure 6.4 – Airflow UI – DAG Graph view

Note how Airflow automatically detects the dependencies and the parallelisms of the tasks. Let's turn on the scheduler for this DAG and see the results of the execution (*Figure 6.5*).

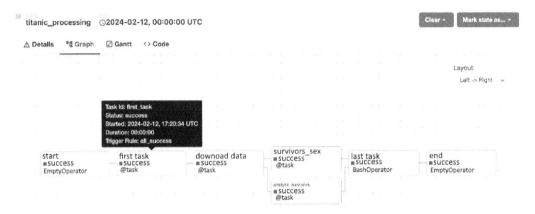

Figure 6.5 – Airflow UI – DAG Graph view after execution

When we turn on the scheduler, as the scheduler is set to @once, Airflow will automatically start the execution. It marks the tasks as a success when they are done. Click on the task with the name first_task and click on **Logs** to check the output (*Figure 6.6*).

Figure 6.6 – Airflow UI – first_task output

Note that the output we programmed is shown in the logs – "And so, it begins!". You can also check the logs of the other tasks to ensure that everything went as expected. Another important view in Airflow is the Gantt chart. Click on that at the top of the page to visualize how much time was spent on each task (*Figure 6.7*). This is a great tool to check for execution bottlenecks and possibilities for optimization.

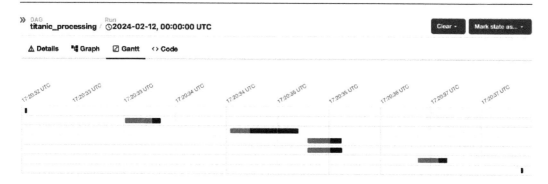

Figure 6.7 – Airflow UI – Gantt view

Congratulations! You just built your first Airflow DAG! Now, let's see how we can integrate Airflow with other tools and orchestrate a more complex pipeline with it.

Airflow integration with other tools

We will take the DAG code developed in the last section and rebuild it with some different tasks. Our DAG is going to download the `Titanic` data, write it to a PostgreSQL table, and write it as a CSV file to Amazon S3. Also, we will create a view with a simple analysis on Postgres directly from Airflow:

1. Create a new Python file in the `dags` folder and name it `postgres_aws_dag.py`. The first part of our code will define the necessary modules. Note that, this time, we are importing the `PostgresOperator` class to interact with this database and the `Variable` class. This will help us manage secrets and parameters in Airflow. We are also creating an SQLAlchemy engine to connect to a local Postgres database and creating an S3 client that will allow writing files to S3:

    ```python
    from airflow.decorators import task, dag
    from airflow.models import Variable
    from airflow.providers.postgres.operators.postgres import
    PostgresOperator
    from datetime import datetime
    import requests
    import pandas as pd
    from sqlalchemy import create_engine
    import boto3

    engine = create_engine('postgresql://postgres:postgres@
    postgres:5432/postgres')

    aws_access_key_id = Variable.get('aws_access_key_id')
    aws_secret_access_key = Variable.get('aws_secret_access_key')
    s3_client = boto3.client(
        's3',
    ```

```
        aws_access_key_id=aws_access_key_id,
        aws_secret_access_key=aws_secret_access_key
)


default_args = {
    'owner': 'Ney',
    'start_date': datetime(2024, 2, 12)
}
```

2. Now, let's start developing our DAG. First, let's define four tasks – one to download the data and the second one to write it as a Postgres table. The third task will create a view in Postgres with a grouped summarization, and the last one will upload the CSV file to S3. This last one is an example of a good practice, a task sending data processing to run outside of Airflow:

```
@dag(
    default_args=default_args,
    schedule_interval="@once",
    description="Insert Data into PostgreSQL and AWS",
    catchup=False,
    tags=['postgres', 'aws']
)
def postgres_aws_dag():

    @task
    def download_data():
        destination = "/tmp/titanic.csv"
        response = requests.get(
"https://raw.githubusercontent.com/neylsoncrepalde/titanic_data_
with_semicolon/main/titanic.csv",
stream=True
        )
        with open(destination, mode="wb") as file:
            file.write(response.content)
        return destination

    @task
    def write_to_postgres(source):
        df = pd.read_csv(source, sep=";")
        df.to_sql('titanic', engine, if_exists="replace",
chunksize=1000, method='multi')

        create_view = PostgresOperator(
        task_id="create_view",
        postgres_conn_id='postgres',
```

```
        sql='''
CREATE OR REPLACE VIEW titanic_count_survivors AS
SELECT
"Sex",
SUM("Survived") as survivors_count
FROM titanic
GROUP BY "Sex"
""",
    )

    @task
    def upload_to_s3(source):
        s3_client.upload_file(source, ' bdok-<ACCOUNT_NUMBER>
', 'titanic.csv')
```

At this point, you should have a bucket created in S3 with the name bdok-<YOUR_ACCOUNT_NUMBER>. Appending your account number in a bucket's name is a great way to guarantee its uniqueness.

3. Now, save your file and take a look at the Airflow UI. Note that the DAG is not available, and Airflow shows an error. Expand the error message and you will see that it's complaining about the variables we are trying to get in our code (*Figure 6.8*). They don't exist just yet. Let's create them.

Figure 6.8 – Airflow UI – variables error

4. In the upper menu, click on **Admin** and choose **Variables**. Now, we will create the Airflow environment variables that we need. Copy your AWS secret access key and access key ID and create those variables accordingly. After creating the variables, you can see that the secret access key is hidden in the UI (*Figure 6.9*). Airflow automatically detects secrets and sensitive credentials and hides them in the UI.

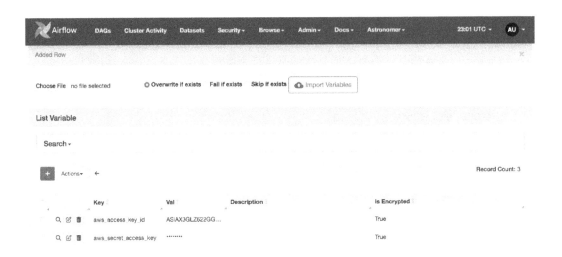

Figure 6.9 – Airflow UI – variables created

5. Now, let's get back to the main page of the UI. Now, the DAG is showing correctly but we are not done yet. For the `PostgresOperator` task to run correctly, it is expecting a Postgres connection (remember the `postgres_conn_id` parameter?). In the upper menu, click on **Admin** and then on **Connections**. Add a new connection as shown in *Figure 6.10*.

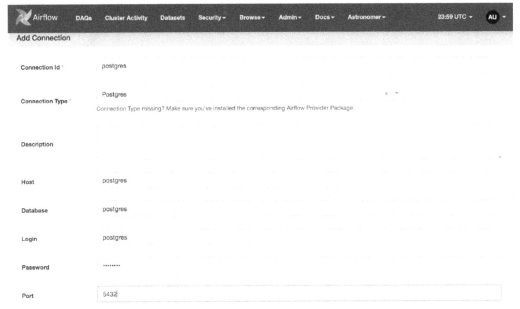

Figure 6.10 – Airflow UI – new connection

6. After creating the connection, let's finish developing our DAG. We have already set the tasks. Now it's time to chain them together:

```
download = download_data()
write = write_to_postgres(download)
write >> create_view
upload = upload_to_s3(download)

execution = postgres_aws_dag()
```

And now, we're good to go. You can also check the complete code, available on GitHub (https://github.com/PacktPublishing/Bigdata-on-Kubernetes/blob/main/Chapter06/dags/postgres_aws_dag.py).

7. Check the DAG's graph in the Airflow UI (*Figure 6.11*) and note how it automatically parallelizes all tasks that are possible – in this case, `write_to_postgres` and `upload_to_s3`. Turn on the DAG's scheduler for it to run. After the DAG runs successfully, check the S3 bucket to validate that the file was correctly uploaded. Then, choose your preferred SQL client, and let's check if the data was correctly ingested to Postgres. For this example, I'm using DBeaver, but you can choose any one you like.

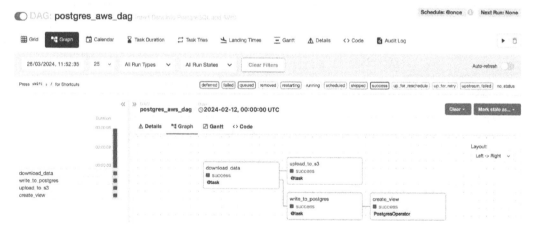

Figure 6.11 – Airflow UI – final DAG

8. In DBeaver, create a new connection to PostgreSQL. For DBeaver's connection, we will use `localhost` as the host for Postgres (in Airflow, we need to use `postgres`, as it is running in a shared network inside Docker). Complete it with the username and password (in this case, `postgres` and `postgres`) and test the connection. If everything is OK, create the connection, and let's run some queries on this database. In *Figure 6.12*, we can see that the data was ingested correctly.

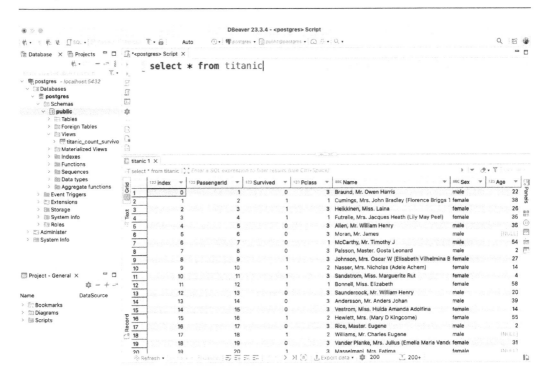

Figure 6.12 – DBeaver – titanic table

9. Now, let's check whether the view was correctly created. The results are shown in *Figure 6.13*.

Figure 6.13 – DBeaver – created view

And that's it! You created a table and a view and uploaded data to AWS using Airflow! To stop Airflow's containers, go back to your terminal and type the following:

```
astro dev kill
```

Summary

In this chapter, we covered the fundamentals of Apache Airflow – from installation to developing data pipelines. You learned how to leverage Airflow to orchestrate complex workflows involving data acquisition, processing, and integration with external systems.

We installed Airflow locally using the Astro CLI and Docker. This provided a quick way to get hands-on without a heavy setup. You were exposed to Airflow's architecture and key components, such as the scheduler, worker, and metadata database. Understanding these pieces is crucial for monitoring and troubleshooting Airflow in production.

Then, there was a major section focused on building your first Airflow DAGs. You used core Airflow operators and the task and DAG decorators to define and chain tasks. We discussed best practices such as keeping tasks small and autonomous. You also learned how Airflow handles task dependencies – allowing parallel execution of independent tasks. These learnings will help you develop effective DAGs that are scalable and reliable.

Later, we integrated Airflow with external tools – writing to PostgreSQL, creating views, and uploading files to S3. This showcased Airflow's versatility to orchestrate workflows involving diverse systems. We also configured Airflow connections and variables to securely pass credentials.

By the end of this chapter, you should have grasped the fundamentals of Airflow and have had hands-on experience building data pipelines. You are now equipped to develop DAGs, integrate other tools, and apply best practices for production-grade workflows. As data teams adopt Airflow, these skills will be invaluable for creating reliable and scalable data pipelines.

In the next chapter, we will study one of the core technologies for real-time data – Apache Kafka.

7

Apache Kafka for Real-Time Events and Data Ingestion

Real-time data and event streaming are crucial components of modern data architectures. By leveraging systems such as Apache Kafka, organizations can ingest, process, and analyze real-time data to drive timely business decisions and actions.

In this chapter, we will cover Kafka's fundamental concepts and architecture that enable it to be a performant, resilient, and scalable messaging system. You will learn how Kafka's publish-subscribe messaging model works with topics, partitions, and brokers. We will demonstrate Kafka setup and configuration, and you will get hands-on experience with producing and consuming messages for topics.

Additionally, you will understand Kafka's distributed and fault-tolerant nature by experimenting with data replication and topic distribution strategies. We will also introduce Kafka Connect for streaming data ingestion from external systems such as databases. You will configure Kafka Connect to stream changes from a SQL database into Kafka topics.

The highlight of this chapter is combining Kafka with Spark Structured Streaming for building real-time data pipelines. You will learn this highly scalable stream processing approach by implementing end-to-end pipelines that consume Kafka topic data, process it using Spark, and write the output into another Kafka topic or external storage system.

By the end of this chapter, you will have gained practical skills to set up Kafka clusters and leverage Kafka's capabilities for building robust real-time data streaming and processing architectures. Companies can greatly benefit from making timely data-driven decisions, and Kafka enables realizing that objective.

In this chapter, we're going to cover the following main topics:

- Getting started with Kafka
- Exploring the Kafka architecture
- Streaming from a database with Kafka Connect
- Real-time data processing with Kafka and Spark

Technical requirements

In this chapter, we will run a Kafka cluster and a Kafka Connect cluster locally using **Docker** and **docker-compose**. It is recommended that you have Docker installed (refer to *Chapter 1* for instructions on that). Usually, `docker-compose` comes with Docker, so no further installation steps are necessary. If you find yourself in the need to manually install `docker-compose`, please refer to `https://docs.docker.com/compose/install/`.

Additionally, we will process data in real time using **Spark**. For installation instructions, please refer to *Chapter 5*.

All the code for this chapter is available online in this book's GitHub repository (`https://github.com/PacktPublishing/Bigdata-on-Kubernetes`) in the `Chapter07` folder.

Getting started with Kafka

Kafka is a popular open source platform for building real-time data pipelines and streaming applications. In this section, we will learn how to get a basic Kafka environment running locally using `docker-compose` so that you can start building Kafka producers and consumers.

`docker-compose` is a tool that helps define and run multi-container Docker applications. With compose, you use a YAML file to configure your application's services then spin everything up with one command. This allows you to avoid having to run and connect containers manually. To run our Kafka cluster, we will define a set of nodes using `docker-compose`. First, create a folder called `multinode` (just to keep our code organized) and create a new file called `docker-compose.yaml`. This is the regular file that `docker-compose` expects to set up the containers (the same as Dockerfile for Docker). To improve readability, we will not show the entire code (it is available at `https://github.com/PacktPublishing/Bigdata-on-Kubernetes/tree/main/Chapter07/multinode`), but a portion of it. Let's take a look:

docker-compose.yaml

```
---
version: '2'
services:
    zookeeper-1:
      image: confluentinc/cp-zookeeper:7.6.0
      environment:
        ZOOKEEPER_SERVER_ID: 1
        ZOOKEEPER_CLIENT_PORT: 22181
        ZOOKEEPER_TICK_TIME: 2000
        ZOOKEEPER_INIT_LIMIT: 5
        ZOOKEEPER_SYNC_LIMIT: 2
        ZOOKEEPER_SERVERS:
```

```
localhost:22888:23888;localhost:32888:33888;localhost:42888:43888
   network_mode: host
   extra_hosts:
     - "mynet:127.0.0.1"


 kafka-1:
   image: confluentinc/cp-kafka:7.6.0
   network_mode: host
   depends_on:
     - zookeeper-1
     - zookeeper-2
     - zookeeper-3
environment:
   KAFKA_BROKER_ID: 1
   KAFKA_ZOOKEEPER_CONNECT:
localhost:22181,localhost:32181,localhost:42181
   KAFKA_ADVERTISED_LISTENERS: PLAINTEXT://localhost:19092
   extra_hosts:
     - "mynet:127.0.0.1"
```

The original Docker Compose file is setting up a Kafka cluster with three Kafka brokers and three Zookeeper nodes (more details on Kafka architecture in the next section). We just left the definition for the first Zookeeper and Kafka brokers as the other ones are the same. Here, we're using Confluent Kafka (an enterprise-ready version of Kafka maintained by Confluent Inc.) and Zookeeper images to create the containers. For the Zookeeper nodes, the key parameters are as follows:

- ZOOKEEPER_SERVER_ID: The unique ID for each Zookeeper server in the ensemble.

- ZOOKEEPER_CLIENT_PORT: The port for clients to connect to this Zookeeper node. We use different ports for each node.

- ZOOKEEPER_TICK_TIME: The basic time unit used by Zookeeper for heartbeats.

- ZOOKEEPER_INIT_LIMIT: The time the Zookeeper servers have to connect to a leader.

- ZOOKEEPER_SYNC_LIMIT: How far out of date a server can be from a leader.

- ZOOKEEPER_SERVERS: Lists all Zookeeper servers in the ensemble in address:leaderElectionPort:followerPort format.

For the Kafka brokers, the key parameters are as follows:

- KAFKA_BROKER_ID: Unique ID for each Kafka broker.

- KAFKA_ZOOKEEPER_CONNECT: Lists the Zookeeper ensemble that Kafka should connect to.

- KAFKA_ADVERTISED_LISTENERS: Advertised listener for external connections to this broker. We use different ports for each broker.

The containers are configured to use host networking mode to simplify networking. The dependencies ensure Kafka only starts after Zookeeper is ready.

This code creates a fully functional Kafka cluster that can handle replication and failures of individual brokers or Zookeepers. Now, we will get those containers up and running. In a terminal, move to the `multinode` folder and type the following:

```
docker-compose up -d
```

This will tell `docker-compose` to get the containers up. If the necessary images are not found locally, they will be automatically downloaded. The `-d` parameter makes `docker-compose` run in **detached** mode. If we don't use this parameter, the terminal will keep printing containers' logs. We don't want that, so we must use `-d`.

To check the logs for one of the Kafka Brokers, run the following command:

```
docker logs multinode-kafka-1-1
```

Here, `multinode-kafka-1-1` is the name of the first Kafka Broker container we defined in the YAML file. With this command, you should be able to visualize Kafka's logs and validate that everything is running correctly. Now, let's take a closer look at Kafka's architecture and understand how it works.

Exploring the Kafka architecture

Kafka has a distributed architecture that consists of brokers, producers, consumers, topics, partitions, and replicas. At a high level, producers publish messages to topics, brokers receive those messages and store them in partitions, and consumers subscribe to topics and process the messages that are published to them.

Kafka relies on an external coordination service called **Zookeeper**, which helps manage the Kafka cluster. Zookeeper helps with controller election – selecting a broker to be the cluster controller. The controller is responsible for administrative operations such as assigning partitions to brokers and monitoring for broker failures. Zookeeper also helps brokers coordinate among themselves for operations such as leader election for partitions.

Kafka **brokers** are the main components of a Kafka cluster and handle all read/write requests from producers/consumers. Brokers receive messages from producers and expose data to consumers. Each broker manages data stored on local disks in the form of partitions. By default, brokers will evenly distribute partitions among themselves. If a broker goes down, Kafka will automatically redistribute those partitions to other brokers. This helps prevent data loss and ensures high availability. Now, let's understand how Kafka handles messages in a **publish-subscribe** (**PubSub**) design and how it guarantees reliability and scalability for the messages' writing and reading.

The PubSub design

Kafka relies on a PubSub messaging pattern to enable real-time data streams. Kafka organizes messages into categories called **topics**. Topics act as feeds or streams of messages. Producers write data to topics and consumers read from topics. For example, a "page-visits" topic would record every visit to a web page. Topics are always multi-producer and multi-subscriber – they can have zero to many producers writing messages to a topic as well as zero to many consumers reading messages. This helps coordinate data streams between applications.

Topics are split into **partitions** for scalability. Each partition acts as an ordered, immutable sequence of messages that is continually appended to. By partitioning topics into multiple partitions, Kafka can scale topic consumption by having multiple consumers reading from a topic in parallel across partitions. Partitions allow Kafka to distribute load horizontally across brokers and allow for parallelism. Data is kept in the order it was produced within each partition.

Kafka provides redundancy and fault tolerance by **replicating partitions** across a configurable number of brokers. A partition will have one broker designated as the "leader" and zero or more brokers acting as "followers." All reads/writes go to the leader. Followers passively replicate the leader by having identical copies of the leader's data. If the leader fails, one of the followers will automatically become the new leader.

Having **replicas** across brokers ensures fault tolerance since data is still available for consumption, even if some brokers go down. The replication factor controls the number of replicas. For example, a replication factor of three means there will be two followers replicating the one leader partition. Common production settings have a minimum of three brokers with a replication factor of two or three.

Consumers label themselves with **consumer group** names, and each record that's published to a topic is only delivered to one consumer in a group. If there are multiple consumers in a group, Kafka will load balance messages across the consumers. Kafka guarantees an ordered, at-least-once delivery of messages within a partition to a single consumer. Consumer groups allow you to scale out consumers while still providing message-ordering guarantees.

Producers publish records to topic partitions. If there is only one partition, all messages will go there. With multiple partitions, producers can choose to either publish randomly across partitions or ensure ordering by using the same partition. This ordering guarantee only applies within a partition, not across partitions.

Producers batch together messages for efficiency and durability. Messages are buffered locally and compressed before being sent to brokers. This batching provides better efficiency and throughput. Producers can choose to wait until a batch is full, or flush based on time or message size thresholds. Producers also replicate data by having it acknowledged by all in-sync replicas before confirming a write. Producers can choose different acknowledgment guarantees, ranging from committing as soon as the leader writes the record or waiting until all followers have replicated.

Consumers read records by subscribing to Kafka topics. Consumer instances can be in separate processes or servers. Consumers pull data from brokers by periodically sending requests for data. Consumers keep track of their position ("offsets") within each partition to start reading from the correct place in case of failures. Consumers typically commit offsets periodically. Offsets are also used to allow consumers to rewind or skip ahead if needed. How exactly do those offsets work? Let's try to understand them a little deeper.

Kafka stores streams of records in categories called topics. Within a topic, records are organized into partitions, which allow for parallel processing and scalability. Each record within a partition gets an *incremental ID number* called an **offset** that uniquely identifies the record within that partition. This offset reflects the order of records within a partition. For example, an offset of three means it's the third record.

When a Kafka consumer reads records from a partition, it keeps track of the offset of the last record it has read. This allows the consumer to only read newer records it hasn't processed yet. If the consumer disconnects and reconnects later, it will start reading again from the last committed offset. The offsets commit log is stored in a Kafka topic named `__consumer_offsets`. This provides durability and allows consumers to transparently pick up where they left off in case of failures.

Offsets enable multiple consumers to read from the same partition while ensuring each record is processed only once by each consumer. The consumers can read at their own pace without interfering with each other. This is a key design feature that enables Kafka's scalability. All these features, when used together, allow Kafka to deliver **exactly-once semantics**. Let's take a closer look at this concept.

How Kafka delivers exactly-once semantics

When processing data streams, three types of semantics are relevant when considering guarantees on data delivery:

- **At-least-once semantics**: In this case, each record in the data stream is guaranteed to be processed at least once but may be processed more than once. This can happen if there is a failure downstream from the data source before the processing is acknowledged. When the system recovers, the data source will resend the unacknowledged data, causing duplicate processing.

- **At-most-once semantics**: In this case, each record will either be processed once or not at all. This prevents duplicate processing but means that in the event of failure, some records may be lost entirely.

- **Exactly-once semantics**: This case combines the guarantees of the other two and ensures each record is processed one and only one time. This is difficult to achieve in practice because it requires coordination between storage and processing to ensure no duplicates are introduced during retries.

Kafka provides a way to enable exactly-once semantics for event processing through a combination of architectural design and integration with stream processing systems. Kafka topics are divided into partitions, which enables data parallelism by spreading the load across brokers. Events with the same key go to the same partition, enabling processing ordering guarantees. Kafka assigns each partition a sequential ID called the offset, which uniquely identifies each event within a partition.

Consumers track their position per partition by storing the offset of the last processed event. If a consumer fails and restarts, it will resume from the last committed offset, ensuring events are not missed or processed twice.

By tightly integrating offset tracking and event delivery with stream processors through Kafka's APIs, Kafka's infrastructure provides the backbone for building exactly-once real-time data pipelines.

Figure 7.1 shows a visual representation of Kafka's architecture:

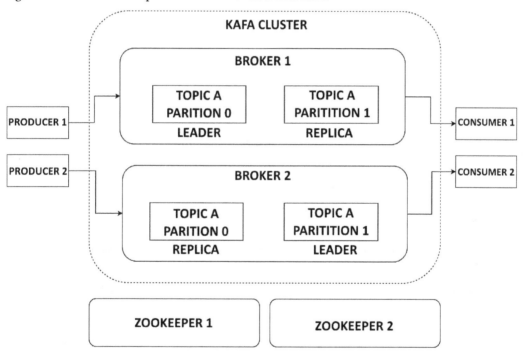

Figure 7.1- Kafka's architecture

Next, we will do a quick exercise to get started and see Kafka in action.

First producer and consumer

After setting up Kafka with `docker-compose`, we have to create a topic that will hold our events. We can do this from outside the container or we can get into the container and run the commands from there. For this exercise, we will access the container and run commands from the inside for didactic purposes. Later in this book, we will study the other option, which can be very handy, especially when Kafka is running on Kubernetes. Let's get started:

1. First, check if all containers are up and running. In a terminal, run the following command:

    ```
    docker-compose ps
    ```

    You should see an output that specifies containers' names, images, commands, and more. Everything seems to be running fine. Note the name of the first Kafka broker container.

2. We will need it to run commands in Kafka from inside the container. To get in, run the following command:

    ```
    CONTAINER_NAME=multinode-kafka-1-1
    docker exec -it $CONTAINER_NAME bash
    ```

    Here, we're creating an environment variable with the first container name (in my case, `multinode_kafka-1_1`) and running the `docker exec` command with the `-it` parameter.

3. Now, we are in the container. Let's declare three environment variables that will help us manage Kafka:

    ```
    BOOTSTRAP_SERVER=localhost:19092
    TOPIC=mytopic
    GROUP=mygroup
    ```

4. Now, we will use the Kafka CLI to create a topic with the `kafka-topics --create` command. Run the following code:

    ```
    kafka-topics --create --bootstrap-server $BOOTSTRAP_SERVER
    --replication-factor 3 --partitions 3 --topic $TOPIC
    ```

    This will create a topic named `mytopic` with a replication factor of 3 (three replicas) and 3 partitions (note that your maximum number of partitions is the number of brokers you have).

5. Although we have a confirmation message in the terminal, it's good to list all the topics inside a cluster:

    ```
    kafka-topics --list --bootstrap-server $BOOTSTRAP_SERVER
    ```

    You should see `mytopic` as output on the screen.

6. Next, let's get some information about our topic:

    ```
    kafka-topics --bootstrap-server $BOOTSTRAP_SERVER --describe
    --topic $TOPIC
    ```

This yields the following output (formatted for better visualization):

```
Topic: mytopic
TopicId: UFt3FOyVRZyYU7TYT1TrsQ
PartitionCount: 3
ReplicationFactor: 3
Configs:
Topic:mytopic Partition:0 Leader:2 Replicas:2,3,1
Topic:mytopic Partition:1 Leader:3 Replicas:3,1,2
Topic:mytopic Partition:2 Leader:1 Replicas:1,2,3
```

This topic structure partitioned across all three brokers and with replications of each partition in all other brokers is exactly what we have seen in *Figure 7.1*.

7. Now, let's build a simple producer and start sending some messages to this topic. In your terminal, type the following:

```
kafka-console-producer --broker-list $BOOTSTRAP_SERVER --topic
$TOPIC
```

This starts a simple console producer.

8. Now, type some messages in the console; they'll be sent to the topic:

```
abc
def
ghi
jkl
mno
pqr
stu
vwx
yza
```

You can type whatever you want.

9. Now, open a different terminal and (preferably) put it beside the first terminal that's running the console producer. We must log in to the container the same way we did in the first terminal:

```
CONTAINER_NAME=multinode-kafka-1-1
docker exec -it $CONTAINER_NAME bash
```

10. Then, create the same necessary environment variables:

```
BOOTSTRAP_SERVER=localhost:19092
TOPIC=mytopic
```

11. Now, we will start a simple console consumer. We will tell this consumer to read all the messages in the topic from the beginning (just for this exercise – this is not recommended for topics in production with a huge amount of data in them). In the second terminal, run the following command:

```
kafka-console-consumer --bootstrap-server $BOOTSTRAP_SERVER
--topic $TOPIC --from-beginning
```

You should see all the messages typed on the screen. Note that they are in a different order because Kafka only keeps messages in order inside partitions.

Across partitions, ordering is not possible (unless you have date-time information inside the message, of course). Press *Ctrl + C* to stop the consumer. You can also press *Ctrl + C* in the producer terminal to stop it. Type exit in both terminals to exit the containers and stop and kill all the containers by running the following command:

```
docker-compose down
```

You can check that all containers were successfully removed with the following command:

```
docker ps -a
```

Now, let's try something different. One of the most common cases of using Kafka is migrating data in real time from tables in a database. Let's see how we can do that simplistically using Kafka Connect.

Streaming from a database with Kafka Connect

In this section, we will read all data that is generated in a Postgres table in real time with Kafka Connect. First, it is necessary to build a custom image of Kafka Connect that can connect to Postgres. Follow these steps:

1. Let's create a different folder for this new exercise. First, create a folder named connect and another folder inside it named kafka-connect-custom-image. Inside the custom image folder, we will create a new Dockerfile with the following content:

```
FROM confluentinc/cp-kafka-connect-base:7.6.0

RUN confluent-hub install --no-prompt confluentinc/kafka-
connect-jdbc:10.7.5 \
&& confluent-hub install --no-prompt confluentinc/kafka-
connect-s3:10.5.8
```

This Docker file bases itself on the confluent Kafka Connect image and installs two connectors – a JDBC source/sink connector and a sink connector for Amazon S3. The former is necessary to connect to a database while the latter will be very handy for delivering events to S3.

2. Build your image with the following commands:

```
cd connect
cd kafka-connect-custom-image
docker build -t connect-custom:1.0.0 .
cd ..
```

Now, in the `connect` folder, you should have a `.env_kafka_connect` file to store your AWS credentials. Remember that credentials should *never* be hardcoded in any configuration files or code. Your `.env_kafka_connect` file should look like this:

```
AWS_DEFAULT_REGION='us-east-1'
AWS_ACCESS_KEY_ID='xxxxxxxxxxxxxxxxxxxxxxxxxxxxxxxxx'
AWS_SECRET_ACCESS_KEY='xxxxxxxxxxxx'
```

3. Save it in the `connect` folder. Then, create a new `docker-compose.yaml` file. The content for this file is available in this book's GitHub repository: `https://github.com/PacktPublishing/Bigdata-on-Kubernetes/blob/main/Chapter07/connect/docker-compose.yaml`.

This Docker Compose file sets up an environment for Kafka and Kafka Connect along with a Postgres database instance. It defines the following services:

*   `zookeeper`: This runs a Zookeeper instance, which Kafka relies on for coordination between nodes. It sets up some configurations, such as port, tick time, and a client port.

*   `broker`: This runs a Kafka broker that depends on the Zookeeper service (a broker cannot exist until the Zookeepers are all up). It configures things such as the broker ID, what Zookeeper instance to connect to, listeners for external connections on ports `9092` and `29092`, replication settings for internal topics Kafka needs, and some performance tuning.

*   `schema-registry`: This runs Confluent Schema Registry, which allows us to store schemas for topics. It depends on the Kafka broker and sets the URL for the Kafka cluster as well as what port to listen on for API requests.

*   `connect`: This runs our customized image of Confluent Kafka Connect. It depends on both the Kafka broker and Schema Registry and sets up bootstrap servers, the group ID, internal topics for storing connector configurations, offsets and status, key-value converters for serialization, Schema Registry integration, and the plugin path for finding more connectors.

*   `rest-proxy`: The runs the Confluent REST proxy, which provides a REST interface to Kafka. It sets up the Kafka broker connection information and Schema Registry.

*   `postgres`: This runs a Postgres database instance that's exposed on port `5432` with some basic credentials set. Note that we are saving the database password in plain text in our code. This should *never* be done in a production environment since it is a security breach. We are only defining the password in this way for local testing.

There is also a custom network defined called `proxynet` that all these services join. This allows inter-service communication by hostname instead of exposing all services to the host machine network.

4. To get these containers up and running, run the following command:

```
docker-compose up -d
```

All the containers should be up in a few minutes.

5. Now, we will continuously insert some simulated data into our Postgres database. To do that, create a new Python file named `make_fake_data.py`. The code is available at `https://github.com/PacktPublishing/Bigdata-on-Kubernetes/tree/main/Chapter07/connect/simulations` folder. This code generates fake data for customers (such as name, address, profession, and email) and inserts it into a database. For it to work, you should have the `faker`, `pandas`, `psycopg2-binary`, and `sqlalchemy` libraries installed. Make sure you install them with `pip install` before running the code. A `requirements.txt` file, along with the code, is provided in this book's GitHub repository.

6. Now, to run the simulations, in a terminal, type the following:

```
python make_fake_data.py
```

This will print the parameters for the simulation (interval of generation, sample size, and the connection string) on the screen and start printing the simulated data. After a few simulations, you can stop it by pressing *Ctrl + C*. Then, use your preferred SQL client (DBeaver is one option) to check if the data was correctly ingested in the database. Run a simple SQL statement (`select * from customers`) to see the data printed in the SQL client.

7. Now, we will register a source JDBC connector to pull data from Postgres. This connector will run as a Kafka Connect process that establishes a JDBC connection to the source database. It uses this connection to execute SQL queries that select data from specific tables. The connector translates the result sets into JSON documents and publishes them to configured Kafka topics. Each table has a dedicated topic created for it. The query that extracts data can be either a simple `SELECT` statement or an incremental query based on timestamp or numeric columns. This allows us to capture new or updated rows.

8. First, we will define a configuration file to deploy the connector on Kafka Connect. Create a folder named `connectors` and a new file named `connect_jdbc_pg_json.config`. The configuration code is shown here:

connect_jdbc_pg_json.config

```
{
    "name": "pg-connector-json",
    "config": {
        "connector.class": "io.confluent.connect.jdbc.
JdbcSourceConnector",
```

```
        "value.converter": "org.apache.kafka.connect.json.
JsonConverter",
        "value.converter.schemas.enable": "true",
        "tasks.max": 1,
        "connection.url": "jdbc:postgresql://postgres:5432/
postgres",
        «connection.user»: «postgres»,
        «connection.password»: «postgres»,
        «mode»: «timestamp»,
        "timestamp.column.name": "dt_update",
        "table.whitelist": "public.customers",
        "topic.prefix": "json-",
        "validate.non.null": "false",
        "poll.interval.ms": 500
    }
}
```

This configuration creates a Kafka connector that will sync rows from the customers table to JSON-formatted Kafka topics, based on timestamp changes to the rows. Let's take a closer look at the parameters that were used:

- name: Names the connector for management purposes.

- connector.class: Specifies the JDBC connector class from Confluent to use.

- value.converter: Specifies that data will be converted into JSON format in Kafka.

- value.converter.schemas.enable: Enables schemas to be stored with the JSON data.

- tasks.max: Limits to one task. This parameter can be increased in a production environment for scalability, depending on the number of partitions in the topic.

- connection.url: Connects to a local PostgreSQL database on port 5432.

- connection.user/.password: PostgreSQL credentials (only in plaintext here for this exercise. Credentials should *never* be hardcoded).

- mode: Specifies to use a timestamp column to detect new/changed rows. You could also use an id column.

- timestamp.column.name: Looks at the dt_update column.

- table.whitelist: Specifies to sync the customers table.

- topic.prefix: Output topics will be prefixed with json-.

- validate.non.null: Allows syncing rows with null values.

- poll.interval.ms: Check for new data every 500ms.

9. Now, we will create a Kafka topic to store the data from the Postgres table. In a terminal, type the following:

```
docker-compose exec broker kafka-topics --create --bootstrap-
server localhost:9092 --partitions 2 --replication-factor 1
--topic json-customers
```

Note that we are using the `docker-compose` API to execute a command inside a container. The first part of the command (`docker-compose exec broker`) tells Docker that we want to execute something in the `broker` service defined in the `docker-compose.yaml` file. The rest of the command is executed inside the broker. We are creating a topic called `json-customers` with two partitions and a replication factor of one (one replica per partition). You should see a confirmation message in the terminal that the topic was created.

10. Next, we will register the connector using a simple API call to Kafka Connect. We will use the `curl` library to do that. In your terminal, type the following:

```
curl -X POST -H "Content-Type: application/json" --data @
connectors/connect_jdbc_pg_json.config localhost:8083/connectors
```

You should see a JSON output with some information about the connector. We can also check if the connector was successfully registered with this call:

```
curl localhost:8083/connectors
```

The name of the connector should be printed in the terminal.

11. Now, do a quick check on the Connect instance logs:

```
docker logs connect
```

Roll up a few lines; you should see the output for the connector registration.

12. Now, let's try a simple console consumer just to validate that the messages are already being migrated to the topic:

```
docker exec -it broker bash
kafka-console-consumer --bootstrap-server localhost:9092 --topic
json-customers --from-beginning
```

You should see the messages in JSON format printed on the screen. Press *Ctrl + C* to stop the consumer and type `exit` to exit the container.

13. Now, we will configure a sink connector to deliver those messages to Amazon S3. First, go to AWS and create a new S3 bucket. S3 bucket names must be unique across all AWS. This way, I recommend setting it with the account name as a suffix (for instance, `kafka-messages-xxxxxxxx`).

Inside the `connectors` folder, create a new file named `connect_s3_sink.config`:

connect_s3_sink.config

```
{
    "name": "customers-s3-sink",
    "config": {
        "connector.class": "io.confluent.connect.
s3.S3SinkConnector",
        "format.class": "io.confluent.connect.s3.format.json.
JsonFormat",
        "keys.format.class": "io.confluent.connect.s3.format.
json.JsonFormat",
        "key.converter": "org.apache.kafka.connect.json.
JsonConverter",
        "value.converter": "org.apache.kafka.connect.json.
JsonConverter",
        "key.converter.schemas.enable": false,
        "value.converter.schemas.enable": false,
        "flush.size": 1,
        "schema.compatibility": "FULL",
        "s3.bucket.name": "<YOUR_BUCKET_NAME>",
        "s3.region": "us-east-1",
        "s3.object.tagging": true,
        "s3.ssea.name": "AES256",
        "topics.dir": "raw-data/kafka",
        "storage.class": "io.confluent.connect.s3.storage.
S3Storage",
        "tasks.max": 1,
        "topics": "json-customers"
    }
}
```

Let's become familiar with the parameters of this connector:

- `connector.class`: Specifies the connector class to use. In this case, it is the Confluent S3 sink connector.

- `format.class`: Specifies the format to use when writing data to S3. Here, we're using `JsonFormat` so that data will be stored in JSON format.

- `key.converter` and `value.converter`: Specify the converter classes to use for serializing the key and values to JSON, respectively.

- `key.converter.schemas.enable` and `value.converter.schemas.enable`: Disable schema validation for keys and values.

- `flush.size`: Specifies the number of records the connector should wait before performing a flush to S3. Here, this parameter is set to 1. However, in production, when you have a large message throughput, it is best to set this value higher so that more messages get delivered to S3 in a single file.

- `schema.compatibility`: Specifies the schema compatibility rule to use. Here, `FULL` means that schemas must be fully compatible.

- `s3.bucket.name`: The name of the S3 bucket to write data to.

- `s3.region`: The AWS region where the S3 bucket is located.

- `s3.object.tagging`: Enables S3 object tagging.

- `s3.ssea.name`: The server-side encryption algorithm to use (AES256, S3 managed encryption, in this case).

- `topics.dir`: Specifies the directory in the S3 bucket to write data to.

- `storage.class`: Specifies the underlying storage class.

- `tasks.max`: The maximum number of tasks for this connector. This should typically be 1 for a sink.

- `topics`: A comma-separated list of topics to get data from to write to S3.

14. Now, we can register the sink connector. In your terminal, type the following:

```
curl -X POST -H "Content-Type: application/json" --data @
connectors/connect_s3_sink.config localhost:8083/connectors
```

Check the logs with `docker logs connect` to validate that the connector was correctly registered and there were no errors in its deployment.

And that's it! You can check the S3 bucket on AWS and see the JSON files coming through. If you want, run the `make_fake_data.py` simulator once again to see more messages be delivered to S3.

Now that you know how to set up a real-time message delivery pipeline, let's introduce some real-time processing in it with Apache Spark.

Real-time data processing with Kafka and Spark

An extremely important part of real-time data pipelines relates to real-time processing. As data gets generated continuously from various sources, such as user activity logs, IoT sensors, and more, we need to be able to make transformations on these streams of data in real time.

Apache Spark's Structured Streaming module provides a high-level API for processing real-time data streams. It builds on top of Spark SQL and provides expressive stream processing using SQL-like operations. Spark Structured Streaming processes data streams using a micro-batch processing model. In this model, streaming data is received and collected into small batches that are processed very quickly, typically within milliseconds. This provides low processing latency while retaining the scalability of batch processing.

We will take from the real-time pipeline that we started with Kafka and build real-time processing on top of it. We will use the Spark Structured Streaming module for that. Create a new folder called `processing` and a file inside it called `consume_from_kafka.py`. The Spark code that processes the data and aggregates the results has been provided here.

The code is also available in this book's GitHub repository: `https://github.com/PacktPublishing/Bigdata-on-Kubernetes/blob/main/Chapter07/connect/processing/consume_from_kafka.py`. This Spark Structured Streaming application is reading from the `json-customers` Kafka topic, transforming the JSON data, and computing aggregations on it before printing the output to the console:

consume_from_kafka.py

```python
from pyspark.sql import SparkSession
from pyspark.sql import functions as f
from pyspark.sql.types import *

spark = (
    SparkSession.builder
    .config("spark.jars.packages", "org.apache.spark:spark-sql-
kafka-0-10_2.12:3.1.2")
    .appName("ConsumeFromKafka")
    .getOrCreate()
)

spark.sparkContext.setLogLevel('ERROR')

df = (
    spark.readStream
    .format('kafka')
    .option("kafka.bootstrap.servers", "localhost:9092")
    .option("subscribe", "json-customers")
    .option("startingOffsets", "earliest")
    .load()
)
```

First, `SparkSession` is created and configured to use the Kafka connector package. Error logging is set to reduce noise and facilitate output visualization in the terminal. Next, a DataFrame is created by reading from the `json-customers` topic using the `kafka` source. It connects to Kafka running on localhost, starts reading from the earliest offset, and represents each message payload as a string:

```
schema1 = StructType([
    StructField("schema", StringType(), False),
    StructField("payload", StringType(), False)
])

schema2 = StructType([
    StructField("name", StringType(), False),
    StructField("gender", StringType(), False),
    StructField("phone", StringType(), False),
    StructField("email", StringType(), False),
    StructField("photo", StringType(), False),
    StructField("birthdate", StringType(), False),
    StructField("profession", StringType(), False),
    StructField("dt_update", LongType(), False)
])

o = df.selectExpr("CAST(value AS STRING)")
o2 = o.select(f.from_json(f.col("value"), schema1).alias("data")).
selectExpr("data.payload")
o2 = o2.selectExpr("CAST(payload AS STRING)")
newdf = o2.select(f.from_json(f.col("payload"), schema2).
alias("data")).selectExpr("data.*")
```

This second block defines two schemas – `schema1` captures the nested JSON structure expected in the Kafka payload, with a schema field and payload field. On the other hand, `schema2` defines the actual customer data schema contained in the payload field.

The value string field, representing the raw Kafka message payload, is extracted from the initial DataFrame. This string payload is parsed as JSON using the defined `schema1` to extract just the payload field. The payload string is then parsed again using `schema2` to extract the actual customer data fields into a new DataFrame called `newdf`:

```
query = (
    newdf
    .withColumn("dt_birthdate", f.col("birthdate"))
    .withColumn("today", f.to_date(f.current_timestamp()))
    .withColumn("age", f.round(
f.datediff(f.col("today"), f.col("dt_birthdate"))/365.25, 0)
```

```
    )
    .groupBy("gender")
    .agg(
        f.count(f.lit(1)).alias("count"),
        f.first("dt_birthdate").alias("first_birthdate"),
        f.first("today").alias("first_now"),
        f.round(f.avg("age"), 2).alias("avg_age")
    )
)
```

Now, the transformations occur – the `birthdate` string is cast to `date`, the current date is fetched, and the age is calculated using `datediff`. The data is aggregated by gender to compute the count, earliest birthdate in data, current date, and average age:

```
(
    query
    .writeStream
    .format("console")
    .outputMode("complete")
    .start()
    .awaitTermination()
)
```

Finally, the aggregated DataFrame is written to the console in append output mode using Structured Streaming. This query will run continuously until it's terminated by running *Ctrl + C*.

To run the query, in a terminal, type the following command:

```
spark-submit --packages org.apache.spark:spark-sql-
kafka-0-10_2.12:3.1.2 processing/consume_from_kafka.py
```

You should see the aggregated data in your terminal. Open another terminal and run more simulations by running the following command:

```
python simulations/make_fake_data.py
```

As new simulations are generated and ingested into Postgres, the Kafka connector will automatically pull them to the `json-customers` topic, at which point Spark will pull those messages, calculate the aggregations in real time, and print the results. After a while, you can hit *Ctrl + C* to stop simulations and then again to stop the Spark streaming query.

Congratulations! You ran a real-time data processing pipeline using Kafka and Spark! Remember to clean up the created resources with `docker-compose down`.

Summary

In this chapter, we covered the fundamental concepts and architecture behind Apache Kafka – a popular open source platform for building real-time data pipelines and streaming applications.

You learned how Kafka provides distributed, partitioned, replicated, and fault-tolerant PubSub messaging through its topics and brokers architecture. Through hands-on examples, you gained practical experience with setting up local Kafka clusters using Docker, creating topics, and producing and consuming messages. You understood offsets and consumer groups that enable fault tolerance and parallel consumption from topics.

We introduced Kafka Connect, which allows us to stream data between Kafka and external systems such as databases. You implemented a source connector to ingest changes from a PostgreSQL database into Kafka topics. We also set up a sink connector to deliver the messages from Kafka to object storage in AWS S3 in real time.

The highlight was building an end-to-end streaming pipeline with Kafka and Spark Structured Streaming. You learned how micro-batch processing on streaming data allows low latency while retaining scalability. The example provided showed how to consume messages from Kafka, transform them using Spark, and aggregate the results in real time.

Through these hands-on exercises, you gained practical experience with Kafka's architecture and capabilities for building robust and scalable streaming data pipelines and applications. Companies can greatly benefit from leveraging Kafka to power their real-time data processing needs to drive timely insights and actions.

In the next chapter, we will finally get all the technologies we've studied so far into Kubernetes. You will learn how to deploy Airflow, Spark, and Kafka in Kubernetes and get them ready to build a fully integrated data pipeline.

Part 3:
Connecting It All Together

In this part, you will learn how to deploy and orchestrate the big data tools and technologies covered in the previous chapters on Kubernetes. You will build scripts to deploy Apache Spark, Apache Airflow, and Apache Kafka on a Kubernetes cluster, making them ready for running data processing jobs, orchestrating data pipelines, and handling real-time data ingestion, respectively. Additionally, you will explore data consumption layers, data lake engines such as Trino, and real-time data visualization with Elasticsearch and Kibana, all deployed on Kubernetes. Finally, you will bring everything together by building and deploying two complete data pipelines, one for batch processing and another for real-time processing, on a Kubernetes cluster. The part also covers the deployment of generative AI applications on Kubernetes and provides guidance on where to go next in your Kubernetes and big data journey.

This part contains the following chapters:

- *Chapter 8, Deploying the Big Data Stack on Kubernetes*
- *Chapter 9, Data Consumption Layer*
- *Chapter 10, Building a Big Data Pipeline on Kubernetes*
- *Chapter 11, Generative AI on Kubernetes*
- *Chapter 12, Where To Go From Here*

8

Deploying the Big Data Stack on Kubernetes

In this chapter, we will cover the deployment of key big data technologies – Spark, Airflow, and Kafka – on Kubernetes. As container orchestration and management have become critical for running data workloads efficiently, Kubernetes has emerged as the de facto standard. By the end of this chapter, you will be able to successfully deploy and manage big data stacks on Kubernetes for building robust data pipelines and applications.

We will start by deploying Apache Spark on Kubernetes using the Spark operator. You will learn how to configure and monitor Spark jobs running as Spark applications on your Kubernetes cluster. Being able to run Spark workloads on Kubernetes brings important benefits such as dynamic scaling, versioning, and unified resource management.

Next, we will deploy Apache Airflow on Kubernetes. You will configure Airflow on Kubernetes, link its logs to S3 for easier debugging and monitoring, and set it up to orchestrate data pipelines built using tools such as Spark. Running Airflow on Kubernetes improves reliability, scaling, and resource utilization.

Finally, we will deploy Apache Kafka on Kubernetes, which is critical for streaming data pipelines. Running Kafka on Kubernetes simplifies operations, scaling, and cluster management.

By the end of this chapter, you will have hands-on experience with deploying and managing big data stacks on Kubernetes. This will enable you to build robust, reliable data applications leveraging Kubernetes as your container orchestration platform.

In this chapter, we're going to cover the following main topics:

- Deploying Spark on Kubernetes
- Deploying Airflow on Kubernetes
- Deploying Kafka on Kubernetes

Technical requirements

For the activities in this chapter, you should have an AWS account and `kubectl`, `eksctl`, and `helm` installed. For instructions on how to set up an AWS account and `kubectl` and `eksctl` installation, refer to *Chapter 1*. For `helm` installation instructions, access `https://helm.sh/docs/intro/install/`.

We will also be using the Titanic dataset for our exercises. You can find the version we will use at `https://github.com/neylsoncrepalde/titanic_data_with_semicolon`.

All code in this chapter is available in the GitHub repository at `https://github.com/PacktPublishing/Bigdata-on-Kubernetes`, in the `Chapter08` folder.

Deploying Spark on Kubernetes

To help us deploy resources on Kubernetes, we are going to use **Helm**. Helm is a package manager for Kubernetes that helps install applications and services. Helm uses templates called **Charts**, which package up installation configuration, default settings, dependencies, and more, into an easy-to-deploy bundle.

On the other hand, we have **Operators**. Operators are custom controllers that extend the Kubernetes API to manage applications and their components. They provide a declarative way to create, configure, and manage complex stateful applications on Kubernetes.

Some key benefits of using operators include the following:

- **Simplified application deployment and lifecycle management**: Operators abstract away low-level details and provide high-level abstractions for deploying applications without needing to understand the intricacies of Kubernetes

- **Integration with monitoring tools**: Operators expose custom metrics and logs, enabling integration with monitoring stacks such as Prometheus and Grafana

- **Kubernetes native**: Operators leverage Kubernetes' extensibility and are written specifically for Kubernetes, allowing them to be cloud-agnostic.

Operators extend Kubernetes by creating **Custom Resource Definitions** (**CRDs**) and controllers. A CRD allows you to define a new resource type in Kubernetes. For example, the SparkOperator defines a SparkApplication resource.

The operator then creates a controller that watches for these custom resources and performs actions based on the resource `spec`.

For example, when a SparkApplication resource is created, the SparkOperator controller will do the following:

- Create the driver and executor Pods based on the spec

- Mount storage volumes

- Monitor the status of the application

- Perform logging and monitoring

Now, let's get to it:

1. To start, let's create an AWS EKS cluster using `eksctl`:

```
eksctl create cluster --managed --alb-ingress-access --node-
private-networking --full-ecr-access --name=studycluster
--instance-types=m6i.xlarge --region=us-east-1 --nodes-min=3
--nodes-max=4 --nodegroup-name=ng-studycluster
```

Remember that this line of code takes several minutes to complete. Now, there are some important configurations to give our Kubernetes cluster permission to create volumes on our behalf. For this, we need to install the AWS EBS CSI driver. This is not required for deploying Spark applications, but it will be very important for Airflow deployment.

2. First, we need to associate the IAM OIDC provider with the EKS cluster, which allows IAM roles and users to authenticate to the Kubernetes API. To do that, in the terminal, type the following:

```
eksctl utils associate-iam-oidc-provider --region=us-east-1
--cluster=studycluster --approve
```

3. Next, we will create an IAM service account called `ebs-csi-controller-sa` in the `kube-system` namespace, with the specified IAM role and policy attached. This service account will be used by the EBS CSI driver:

```
eksctl create iamserviceaccount --name ebs-csi-controller-sa
--namespace kube-system --cluster studycluster --role-name
AmazonEKS_EBS_CSI_DriverRole --role-only --attach-policy-arn
arn:aws:iam::aws:policy/service-role/AmazonEBSCSIDriverPolicy
--approve
```

4. Finally, we will enable the EBS CSI driver in the cluster and link it to the service account and role created earlier. Remember to change `<YOUR_ACCOUNT_NUMBER>` for the real value:

```
eksctl create addon --name aws-ebs-csi-driver --cluster
studycluster --service-account-role-arn arn:aws:iam::<YOUR_
ACCOUNT_NUMBER>:role/AmazonEKS_EBS_CSI_DriverRole --force
```

5. Now, let's start the actual Spark operator deployment. We will create a namespace to organize our resources next:

```
kubectl create namespace spark-operator
```

6. Next, we will use the SparkOperator Helm chart available online to deploy the operator:

```
helm install spark-operator https://github.com/kubeflow/spark-
operator/releases/download/spark-operator-chart-1.1.27/spark-
operator-1.1.27.tgz --namespace spark-operator --set webhook.
enable=true
```

7. Check whether the operator was correctly deployed:

```
kubectl get pods -n spark-operator
```

You should see output like this:

```
NAME                                  READY   STATUS
spark-operator-74db6fcf98-grhdw       1/1     Running
spark-operator-webhook-init-mw8gf     0/1     Completed
```

8. Next, we need to register our AWS credentials as a Kubernetes Secret to make them available for Spark. This will allow our Spark applications to access resources in AWS:

```
kubectl create secret generic aws-credentials --from-
literal=aws_access_key_id=<YOUR_ACCESS_KEY_ID> --from-
literal=aws_secret_access_key="<YOUR_SECRET_ACCESS_KEY>" -n
spark-operator
```

9. Now, it's time to develop our Spark code. By now, you should have the Titanic dataset stored on Amazon S3. At `https://github.com/PacktPublishing/Bigdata-on-Kubernetes/blob/main/Chapter08/spark/spark_job.py`, you will find simple code that reads the Titanic dataset from the S3 bucket and writes it into another bucket (this second S3 bucket must have been created previously – you can do it in the AWS console).

10. Save this file as `spark_job.py` and upload it to a different S3 bucket. This is where the SparkOperator is going to look for the code to run the application. Note that this PySpark code is slightly different from what we saw earlier, in *Chapter 5*. Here, we are setting Spark configurations separately from the Spark session. We will go through those configurations in detail:

   - `.set("spark.cores.max", "2")`: This limits the maximum number of cores this Spark application will use to two. This prevents the overallocation of resources.

   - `.set("spark.executor.extraJavaOptions", "-Dcom.amazonaws.services.s3.enableV4=true")` and `.set("spark.driver.extraJavaOptions", "-Dcom.amazonaws.services.s3.enableV4=true")`: These enable support for reading and writing to S3 using Signature Version 4 authentication, which is more secure.

   - `.set("spark.hadoop.fs.s3a.fast.upload", True)`: This property enables the fast upload feature of the S3A connector which improves performance when saving data to S3.

   - `.set("spark.hadoop.fs.s3a.impl", "org.apache.hadoop.fs.s3a.S3AFileSystem")`: This configuration sets the S3 FileSystem implementation to use the newer, optimized `s3a` instead of the older `s3` connector.

- `.set("spark.hadoop.fs.s3a.aws.crendentials.provider", "com.amazonaws.auth.EnvironmentVariablesCredentials")`: This configures Spark to obtain AWS credentials from the environment variables, rather than needing to specify them directly in code.

- `.set("spark.jars.packages", "org.apache.hadoop:hadoop-aws:2.7.3")`: This adds a dependency on the Hadoop AWS module so Spark can access S3.

Also note that, by default, Spark uses log-level `INFO`. In this code, we set it to `WARN` to reduce logging and improve logs' readability. Remember to change `<YOUR_BUCKET>` for your own S3 buckets.

11. After uploading this code to S3, it's time to create a YAML file with the SparkApplication definitions. The content for the code is available at `https://github.com/PacktPublishing/Bigdata-on-Kubernetes/blob/main/Chapter08/spark/spark_job.yaml`.

The code defines a new SparkApplication resource. This is only possible because the SparkOperator created the SparkApplication custom resource. Let's take a closer look at what this YAML definition is doing.

- The first block of the YAML file specifies the apiVersion and the kind of resource as Spark application. It also sets a name and namespace for the application.

- The second block defines a volume mount called "ivy" that will be used to cache dependencies and avoid fetching them for each job run. It mounts to `/tmp` in the driver and executors.

- The third block configures Spark properties, enabling the Ivy cache directory and setting the resource allocation batch size for Kubernetes.

- The fourth block configures Hadoop properties to use the S3A file system implementation.

- The fifth block sets this Spark application as a Python one, the Python version to use, running in cluster mode, and the Docker image to use – in this case, a previously prepared Spark image that integrates with AWS and Kafka. It also defines that the image will always be pulled from Docker Hub, even if it is already present in the cluster.

- The sixth block specifies the main Python application file location in S3 and the Spark version – in this case, 3.1.1.

- The seventh block sets `restartPolicy` to `Never`, so the application runs only once.

The remaining blocks set configuration for the driver and executor Pods. Here, we set up AWS access key secrets for accessing S3, we request one core and 1 GB memory for the driver and the same resources for the executors, we mount an `emptyDir` volume called "ivy" for caching dependencies, and we set Spark and driver Pod labels for tracking.

12. Once this file is saved on your computer and you already have the `.py` file on S3, it's time to run the Spark Application. In the terminal, type the following:

```
kubectl apply -f spark_job.yaml -n spark-operator
```

We can check whether the application was successfully submitted with this:

```
kubectl get sparkapplication -n spark-operator
```

We can get a few more details on the application using the following:

```
kubectl describe sparkapplication/test-spark-job -n spark-operator
```

To see the logs of our Spark Application, type this:

```
kubectl logs test-spark-job-driver -n spark-operator
```

And that's it! You just ran your first Spark application on Kubernetes! Kubernetes won't let you deploy another job with the same name, so, to test again, you should delete the application:

```
kubectl delete sparkapplication/test-spark-job -n spark-operator
```

Now, let's see how to deploy Airflow on Kubernetes using the official Helm chart.

Deploying Airflow on Kubernetes

Airflow deployment on Kubernetes is very straightforward. Nevertheless, there are some important details in the Helm chart configuration that we need to pay attention to.

First, we will download the most recent Helm chart to our local environment:

```
helm repo add apache-airflow https://airflow.apache.org
```

Next, we need to configure a `custom_values.yaml` file to change the default deployment configurations for the chart. An example of this YAML file is available at `https://github.com/PacktPublishing/Bigdata-on-Kubernetes/blob/main/Chapter08/airflow/custom_values.yaml`. We will not go through the entire file but just the most important configurations that are needed for this deployment:

1. In the `defaultAirflowTag` and `airflowVersion` parameters, make sure to set `2.8.3`. This is the latest Airflow version available for the 1.13.1 Helm chart version.

2. The `executor` parameter should be set to `KubernetesExecutor`. This guarantees that Airflow will use Kubernetes infrastructure to launch tasks dynamically.

3. In the `env` section, we will configure "remote logging" to allow Airflow to save logs in S3. This is best practice for auditing and saving Kubernetes storage resources. Here, we configure three environment variables for Airflow. The first one sets remote logging to `"True"`; the second one defines in which S3 bucket and folder Airflow will write logs, and the last one defines a

"connection" that Airflow will use to authenticate in AWS. We will have to set this in the Airflow UI later. This is an example of what this block should look like:

```
env:
    - name: "AIRFLOW__LOGGING__REMOTE_LOGGING"
    value: "True"
    - name: "AIRFLOW__LOGGING__REMOTE_BASE_LOG_FOLDER"
    value: "s3://airflow-logs-<YOUR_ACCOUNT_NUMBER>/airflow-
logs/"
    - name: "AIRFLOW__LOGGING__REMOTE_LOG_CONN_ID"
    value: "aws_conn"
```

4. In the webserver block, we must configure the first user credentials and the type of service. The `service` parameter should be set to "LoadBalancer" so we can access the Airflow UI from a browser. The `defaultUser` block should look like this:

```
defaultUser:
    enabled: true
    role: Admin
    username: <YOUR_USERNAME>
    email: admin@example.com
    firstName: NAME
    lastName: LASTNAME
    password: admin
```

It is important to have a simple password in the values file and change it in the UI as soon as the deployment is ready. This way, your credentials do not get stored in plain text. This would be a major security incident.

5. The `redis.enabled` parameter should be set to `false`. As we are using the Kubernetes executor, Airlfow will not need Redis to manage tasks. If we don't set this parameter to `false`, Helm will deploy a Redis Pod anyway.

6. Lastly, in the `dags` block, we will configure `gitSync`. This is the easiest way to send our DAG files to Airflow and keep them updated in GitHub (or any other Git repository you prefer). First, you should create a GitHub repository with a folder named `"dags"` to store Python DAG files. Then, you should configure the `gitSync` block as follows:

```
gitSync:
    enabled: true

    repo: https://github.com/<USERNAME>/<REPO_NAME>.git
    branch: main
    rev: HEAD
    ref: main
    depth: 1
    maxFailures: 0
    subPath: "dags"
```

Note that we omitted several comments in the original file for readability. The `custom_values.yaml` file is ready for deployment. We can now proceed with this command in the terminal:

```
helm install airflow apache-airflow/airflow --namespace airflow
--create-namespace -f custom_values.yaml
```

This deployment can take a few minutes to finish because Airflow will do a database migration job before making the UI available.

Next, we need to get the URL for the UI's LoadBalancer. In the terminal, type this:

```
kubectl get svc -n airflow
```

In the columns EXTERNAL-IP, you will notice one not empty value for the `airflow-webserver` service. Copy this URL and paste it into your browser, adding "`:8080`" to access Airflow's correct port.

After logging in to the UI, click on **Admin** and **Connections** in the menu to configure an AWS connection. Set the name to `aws_conn` (as we stated in the values file), choose **Amazon Web Services** for the connection type, and enter your access key ID and the secret access key. (At this point, you should have your AWS credentials stored locally – if you don't, in the AWS console, go to IAM and generate new credentials for your user. You will not be able to see the old credentials onscreen.)

Finally, we will use DAG code adapted from *Chapter 5* that will run smoothly on Kubernetes. This DAG will download the Titanic dataset automatically from the internet, perform simple calculations, and print the results, which will be accessed on the "logs" page. The content for the code is available at `https://github.com/PacktPublishing/Bigdata-on-Kubernetes/blob/main/Chapter08/airflow/dags/titanic_dag.py`.

Upload a copy of this Python code to your GitHub repository and, in a few seconds, it will show on the Airflow UI. Now, activate your DAG (click the `switch` button) and follow the execution (*Figure 8.1*).

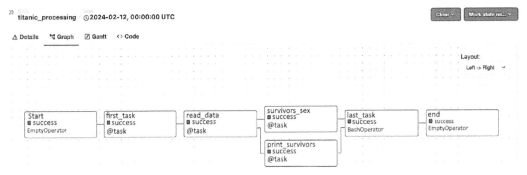

Figure 8.1 – DAG successfully executed

Then click on any task to select it and click on **Logs**. You will see Airflow logs being read directly from S3 (*Figure 8.2*).

Figure 8.2 – Airflow logs in S3

Congratulations! You now have Airflow up and running on Kubernetes. In *Chapter 10*, we will put all these pieces together to build a fully automated data pipeline.

Next, we will deploy a Kafka cluster on Kubernetes using the Strimzi operator. Let's get to it.

Deploying Kafka on Kubernetes

Strimzi is an open source operator that facilitates the deployment and management of Kafka clusters on Kubernetes, creating new CRDs. It is developed and maintained by the Strimzi project, which is part of the **Cloud Native Computing Foundation** (**CNCF**). The Strimzi operator provides a declarative approach to managing Kafka clusters on Kubernetes. Instead of manually creating and configuring Kafka components, you define the desired state of your Kafka cluster using Kubernetes custom resources. The operator then takes care of deploying and managing the Kafka components according to the specified configuration:

1. To deploy Strimzi in Kubernetes, first, we need to install its Helm chart:

    ```
    helm repo add strimzi https://strimzi.io/charts/
    ```

2. Next, we install the operator with the following command:

```
helm install kafka strimzi/strimzi-kafka-operator --namespace
kafka --create-namespace --version 0.40.0
```

3. We can check whether the deployment was successful with the following:

```
helm status kafka -n kafka
kubectl get pods -n kafka
```

4. Now, it's time to configure the deployment of our Kafka cluster. Here is a YAML file to configure a Kafka cluster using the new CRDs. Let's break it into pieces for better understanding:

kafka_jbod.yaml

```
apiVersion: kafka.strimzi.io/v1beta2
kind: Kafka
metadata:
    name: kafka-cluster
spec:
    kafka:
    version: 3.7.0
    replicas: 3
```

The first part of the code specifies the API version and the kind of resource being defined. In this case, it's a Kafka resource managed by the Strimzi operator. Then, we define metadata for the Kafka resource, specifically its name, which is set to `kafka-cluster`. The next block specifies the configuration for the Kafka brokers. We're setting the Kafka version to 3.7.0 and specifying that we want three replicas (Kafka broker instances) in the cluster:

```
listeners:
  - name: plain
    port: 9092
    type: internal
    tls: false
  - name: tls
    port: 9093
    type: internal
    tls: true
  - name: external
    port: 9094
    type: loadbalancer
    tls: false
```

Next, we define the listeners for the Kafka brokers. We're configuring three listeners:

- `plain`: An internal listener on port `9092` without TLS encryption

- `tls`: An internal listener on port `9093` with TLS encryption enabled

- **external**: An external listener on port 9094 exposed as a LoadBalancer service, without TLS encryption

```
readinessProbe:
  initialDelaySeconds: 15
  timeoutSeconds: 5
livenessProbe:
  initialDelaySeconds: 15
  timeoutSeconds: 5
```

The next block configures the readiness and liveness probes for the Kafka brokers. The readiness probe checks whether the broker is ready to accept traffic, while the liveness probe checks whether the broker is still running. The `initialDelaySeconds` parameter specifies the number of seconds to wait before performing the first probe, and `timeoutSeconds` specifies the number of seconds after which the probe is considered failed:

```
config:
  default.replication.factor: 3
  num.partitions: 9
  offsets.topic.replication.factor: 3
  transaction.state.log.replication.factor: 3
  transaction.state.log.min.isr: 1
  log.message.format.version: "3.7"
  inter.broker.protocol.version: "3.7"
  min.insync.replicas: 2
  log.retention.hours: 2160
```

This `kafka.config` block specifies various configuration options for the Kafka brokers, such as the default replication factor, the number of partitions for new topics, the replication factor for the offsets and transaction state log topics, the log message format version, and the log retention period (in hours). The default log retention for Kafka is 7 days (168 hours), but we can change this parameter according to our needs. It is important to remember that longer retention periods imply more disk storage usage, so be careful:

```
storage:
  type: jbod
  volumes:
  - id: 0
    type: persistent-claim
    size: 15Gi
    deleteClaim: false
  - id: 1
    type: persistent-claim
    size: 15Gi
    deleteClaim: false
```

The `kafka.storage` block configures the storage for the Kafka brokers. We're using the **Just a Bunch of Disks (JBOD)** storage type, which allows us to specify multiple persistent volumes for each broker. In this case, we're defining two persistent volume claims of 15 GiB each, with `deleteClaim` set to `false` to prevent the persistent volume claims from being deleted when the Kafka cluster is deleted:

```
resources:
  requests:
    memory: 512Mi
    cpu: "500m"
  limits:
    memory: 1Gi
    cpu: "1000m"
```

Next, the `kafka.resources` block specifies the resource requests and limits for the Kafka brokers. We're requesting 512 MiB of memory and 500 millicpu, and setting the memory limit to 1 GiB and the CPU limit to 1 cpu:

```
zookeeper:
replicas: 3
storage:
    type: persistent-claim
    size: 10Gi
    deleteClaim: false
resources:
  requests:
    memory: 512Mi
    cpu: "250m"
  limits:
    memory: 1Gi
    cpu: "500m"
```

Finally, we have a `zookeeper` block that configures the ZooKeeper ensemble used by the Kafka cluster. We're specifying three replicas for ZooKeeper, using a 10 GiB persistent volume claim for storage, and setting resource requests and limits similar to the Kafka brokers.

5. Once the configuration file is ready on your machine, type the following command to deploy the cluster to Kubernetes:

```
kubectl apply -f kafka_jbod.yaml -n kafka
```

Let's check whether the Kafka cluster was correctly deployed:

```
kubectl get kafka -n kafka
```

This yields the following output:

```
NAME              DESIRED KAFKA REPLICAS    DESIRED ZK REPLICAS
kafka-class   3                             3
```

We can also get detailed information about the deployment:

```
kubectl describe kafka -n kafka
```

Now, check the Pods:

```
kubectl get pods -n kafka
```

The output shows the three Kafka brokers and the ZooKeeper instances:

```
NAME                        READY    STATUS
kafka-class-kafka-0         1/1      Running
kafka-class-kafka-1         1/1      Running
kafka-class-kafka-2         1/1      Running
kafka-class-zookeeper-0     1/1      Running
kafka-class-zookeeper-1     1/1      Running
kafka-class-zookeeper-2     1/1      Running
```

Congrats! You have a fully operational Kafka cluster running on Kubernetes. Now, the next step is to deploy a Kafka Connect cluster and make everything ready for a real-time data pipeline. We will not do that right now for cloud cost efficiency, but we will get back to this configuration in *Chapter 10*.

Summary

In this chapter, you learned how to deploy and manage key big data technologies such as Apache Spark, Apache Airflow, and Apache Kafka on Kubernetes. Deploying these tools on Kubernetes provides several benefits, including simplified operations, better resource utilization, scaling, high availability, and unified cluster management.

You started by deploying the Spark operator on Kubernetes and running a Spark application to process data from Amazon S3. This allows you to leverage Kubernetes for running Spark jobs in a cloud-native way, taking advantage of dynamic resource allocation and scaling.

Next, you deployed Apache Airflow on Kubernetes using the official Helm chart. You configured Airflow to run with the Kubernetes executor, enabling it to dynamically launch tasks on Kubernetes. You also set up remote logging to Amazon S3 for easier monitoring and debugging. Running Airflow on Kubernetes improves reliability, scalability, and resource utilization for orchestrating data pipelines.

Finally, you deployed Apache Kafka on Kubernetes using the Strimzi operator. You configured a Kafka cluster with brokers, a ZooKeeper ensemble, persistent storage volumes, and internal/external listeners. Deploying Kafka on Kubernetes simplifies operations, scaling, high availability, and cluster management for building streaming data pipelines.

Overall, you now have hands-on experience with deploying and managing the key components of a big data stack on Kubernetes. This will enable you to build robust, scalable data applications and pipelines leveraging the power of container orchestration with Kubernetes. The skills learned in this chapter are crucial for running big data workloads efficiently in cloud-native environments.

In the next chapter, we will see how to build a data consumption layer on top of Kubernetes and how to connect those layers with tools to visualize and use data.

9

Data Consumption Layer

In today's data-driven world, organizations are dealing with an ever-increasing volume of data, and the ability to effectively consume and analyze this data is crucial for making informed business decisions. As we delve into the realm of big data on Kubernetes, we must address the critical component of the data consumption layer. This layer serves as the bridge between the vast repositories of data and the business analysts who need to extract valuable insights and make decisions that have an impact on the business.

In this chapter, we will explore two powerful tools that will enable you to unlock the potential of your Kubernetes-based data architecture: **Trino** and **Elasticsearch**. Trino, a distributed SQL query engine, will empower you to directly query your data lake, eliminating the need for a traditional data warehouse. You will learn how to deploy Trino on Kubernetes, monitor its performance, and execute SQL queries against your data stored in Amazon S3.

Furthermore, we will introduce Elasticsearch, a highly scalable and efficient search engine widely used in real-time data pipelines, along with **Kibana**, its powerful data visualization tool. You will gain hands-on experience in deploying Elasticsearch on Kubernetes, indexing data for optimized storage and retrieval, and building simple yet insightful visualizations using Kibana. This combination will equip you with the ability to analyze real-time data streams and uncover valuable patterns and trends.

By the end of this chapter, you will have acquired the skills necessary to successfully deploy and utilize Trino and Elasticsearch on Kubernetes. You will be able to execute SQL queries directly against your data lake, monitor query execution and history, and leverage the power of Elasticsearch and Kibana for real-time data analysis and visualization.

In this chapter, we're going to cover the following main topics:

- Getting started with SQL query engines
- Deploying Trino in Kubernetes
- Deploying Elasticsearch in Kubernetes
- Running queries and connecting to other tools

Technical requirements

For this chapter, you should have an AWS EKS cluster ready for deployment and DBeaver Community locally installed (`https://dbeaver.io/`) We will continue working on the cluster we deployed in *Chapter 8*. All the code for this chapter is available at `https://github.com/PacktPublishing/Bigdata-on-Kubernetes` in the `Chapter09` folder.

Getting started with SQL query engines

In the world of big data, the way we store and analyze data has undergone a significant transformation. Traditional data warehouses, which were once the go-to solution for data analysis, have given way to more modern and scalable approaches, such as SQL query engines. These engines, such as **Trino** (formerly known as Presto), **Dremio**, and **Apache Spark SQL**, offer a high-performance, cost-effective, and flexible alternative to traditional data warehouses.

Next, we are going to see the main differences between data warehouses and SQL query engines.

The limitations of traditional data warehouses

Traditional data warehouses were designed to store and analyze structured data from relational databases. However, with the advent of big data and the proliferation of diverse data sources, such as log files, sensor data, and social media data, the limitations of data warehouses became apparent. These limitations include the following:

- **Scalability**: Data warehouses often struggle to scale horizontally, making it challenging to handle large volumes of data efficiently

- **Data ingestion**: The process of **extracting, transforming, and loading** (ETL) data into a data warehouse can be complex, time-consuming, and resource-intensive

- **Cost**: Data warehouses can be expensive to set up and maintain, especially when dealing with large volumes of data

- **Flexibility**: Data warehouses are typically optimized for structured data and may not handle semi-structured or unstructured data as efficiently

SQL query engines were developed to address these limitations. Let's see how they work.

The rise of SQL query engines

SQL query engines, such as Trino, provide a distributed, scalable, and cost-effective solution for querying large datasets stored in various data sources, including object storage (e.g., Amazon S3, Google Cloud Storage, and Azure Blob Storage), relational databases, and NoSQL databases. We will take a deeper dive into SQL query engines' architecture in the next section.

Here are some key advantages of SQL query engines:

- **High performance**: SQL query engines are designed to leverage the power of distributed computing, allowing them to process large datasets in parallel across multiple nodes. This parallelization enables high-performance queries, even on massive datasets.

- **Cost-effective**: By leveraging object storage and separating storage from compute, SQL query engines can significantly reduce the cost of data storage and processing compared to traditional data warehouses.

- **Scalability**: SQL query engines can scale horizontally by adding more nodes to the cluster, enabling them to handle increasing volumes of data efficiently.

- **Flexibility**: SQL query engines can query a wide range of data sources, including structured, semi-structured, and unstructured data, making them highly flexible and adaptable to various data formats and storage systems.

- **Open source**: Many SQL query engines are open source projects, allowing organizations to leverage the power of community contributions and avoid vendor lock-in.

Now, let's understand the underlying architecture of this type of solution.

The architecture of SQL query engines

SQL query engines such as Trino are designed to work in a distributed computing environment, where multiple nodes work together to process queries and return results. The architecture typically consists of the following components:

- A **coordinator node**, which is responsible for parsing SQL queries, creating a distributed execution plan, and coordinating the execution of the query across the worker nodes.

- A set of **worker nodes**, which are responsible for executing the tasks assigned by the coordinator node. They read data from the underlying data sources, perform computations, and exchange intermediate results with other worker nodes as needed.

- A **metadata store**, which contains information about the data sources, table definitions, and other metadata required for query execution.

When a user submits a SQL query to the SQL query engine, this is what occurs:

1. First, the coordinator node receives the query and parses it to create a distributed **execution plan**.

2. The execution plan is divided into smaller tasks, and these tasks are assigned to the available worker nodes.

3. The worker nodes read data from the underlying data sources, perform computations, and exchange intermediate results as needed.

4. The coordinator node collects and combines the results from the worker nodes to produce the final query result, which is returned to the user of the client application.

This distributed architecture allows SQL query engines to leverage the combined computing power of multiple nodes, enabling them to process large datasets efficiently and deliver high-performance query execution.

In the case of Trino, it can directly connect to object storage systems such as Amazon S3, Azure Blob Storage, or Google Cloud Storage, where data is often stored in formats such as Parquet, ORC, or CSV. Trino can read and process this data directly from the object storage, without the need for intermediate data loading or transformation steps. This capability eliminates the need for a separate data ingestion process, reducing complexity and enabling faster time to insight.

Trino's distributed architecture allows it to split the query execution across multiple worker nodes, each processing a portion of the data in parallel. This parallelization enables Trino to leverage the combined computing power of the cluster, resulting in high-performance query execution, even on massive datasets.

Furthermore, Trino's cost-effectiveness stems from its ability to separate storage from compute. By leveraging object storage for data storage, organizations can take advantage of the low-cost and scalable nature of these storage systems, while dynamically provisioning compute resources (worker nodes) as needed for query execution. This separation of concerns allows organizations to optimize their infrastructure costs and scale resources independently based on their specific needs.

Now, let's move on to a hands-on exercise and see how we can deploy Trino to Kubernetes and connect it to Amazon S3 as a data source.

Deploying Trino in Kubernetes

Trino deployment is very straightforward using its official Helm chart. First, we install the chart with the following:

```
helm repo add trino https://trinodb.github.io/charts
```

Next, we will configure the custom_values.yaml file. The full version of the file is available at https://github.com/PacktPublishing/Bigdata-on-Kubernetes/blob/main/Chapter09/trino/custom_values.yaml. There are only a few custom configurations needed for this deployment. First, the server.workers parameter allows us to set the number of worker pods we want for the Trino cluster. We will set this to 2 but it is advisable to scale if you will run queries on big data:

```
server:
  workers: 2
```

In the block of parameters, set the image.tag parameter to 432 as this is the latest Trino version compatible with the chart version we are using (0.19.0):

```
image:
  registry: ""
```

```
repository: trinodb/trino
tag: 432
```

In the `additionalCatalogs` section, we must configure Trino to use the AWS Glue Data Catalog. The block should be as shown here:

```
additionalCatalogs:
  hive: |
    connector.name=hive
    hive.metastore=glue
```

Finally, we will set the `service.type` parameter to `LoadBalancer` to be able to access Trino from outside AWS (for testing only, not suited for production):

```
service:
  type: LoadBalancer
  port: 8080
```

And that's it. We are ready to launch Trino on Kubernetes.

> **Note**
>
> We are not using any authentication method (password, OAuth, certificate, etc.). In a production environment, you should set an appropriate authentication method and keep the traffic to Trino private inside your VPC (private network), not exposing the load balancer to the internet as we are doing here. This simple configuration is just for training and non-critical data.

After saving the `custom_values.yaml` file, use the following command to deploy Trino:

```
helm install trino trino/trino -f custom_values.yaml -n trino
--create-namespace --version 0.19.0
```

Now, we can check whether the deployment was successful with the following:

```
kubectl get pods,svc -n trino
```

This yields this output:

```
NAME                                  READY   STATUS
pod/trino-coordinator-dbbbcf9d9-94wsn  1/1     Running
pod/trino-worker-6c58b678cc-6fgjs      1/1     Running
pod/trino-worker-6c58b678cc-hldrb      1/1     Running


NAME           TYPE           CLUSTER-IP       EXTERNAL-IP
service/trino  LoadBalancer   10.100.246.148   xxxx.us-east-1.elb.
amazonaws.com
```

The output was simplified to improve visualization. We can see one coordinator node and two workers, which is what we set. Now, copy the URL provided in the `EXTERNAL-IP` column of the output and paste it into your browser, adding `:8080` at the end. You should see a login page.

Figure 9.1 – Trino login page

The default user is `trino`. No password is required as we did not set any in the deployment. After clicking **Log In**, you will see Trino's monitoring page.

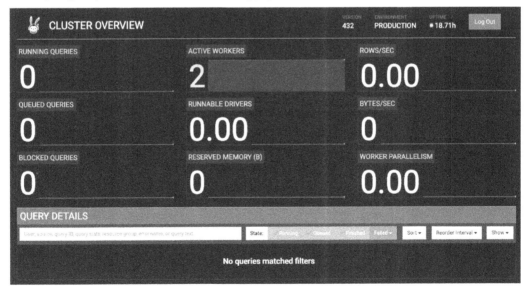

Figure 9.2 – Trino's monitoring page

Next, we will use the same `LoadBalancer` URL to interact with Trino using DBeaver, an open source SQL client.

Connecting DBeaver with Trino

To connect with Trino, first, open DBeaver and create a new Trino connection. In the configuration section (**General**), insert the URL in the **Host** space and `trino` as the username. Leave the password blank.

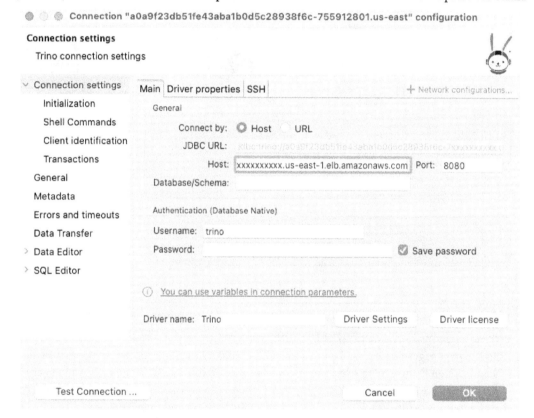

Figure 9.3 – DBeaver connection configuration

Then, click on **Test Connection …**. If this is the first time you are configuring a Trino connection, DBeaver will automatically find the necessary drivers and show a new window asking you to download it. You can hit **OK** and go through the installation, then finish the configuration.

Before we try to access our data, we need to catalog some data and make it available in Glue Data Catalog, as well as setting up an IAM permission that will allow Trino to access the catalog and the underlying data. Let's get to it.

Download the dataset from `https://github.com/neylsoncrepalde/titanic_data_with_semicolon` and store the CSV file in an S3 bucket inside a folder named `titanic`. Glue only understands tables from folders, not from isolated files. Now, we will create a **Glue crawler**. This crawler will look into the dataset, map its columns and column types, and register the metadata in the catalog:

1. In your AWS account, type `Glue` to enter the AWS Glue service, expand the **Data Catalog** option in the side menu, and click on **Crawlers** (*Figure 9.4*).

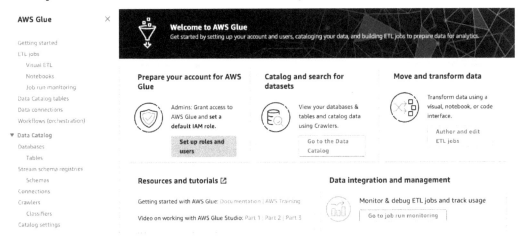

Figure 9.4 – AWS Glue landing page

2. Next, click on **Create crawler** and start filling in the information. In the **Name** section, type `bdok-titanic-crawler` (you can choose any name). Click **Next**.

3. On the next page, click on **Add a data source** and make sure **S3** is selected in the first field. Then, click on **Browse S3** to select the folder where the Titanic dataset is stored. The final path should look like this: `s3://<YOUR_BUCKET_NAME>/titanic/`. You can leave the other configurations as the default. Click on **Add an S3 data source** and then **Next**.

4. In the next step, click on **Create new IAM role** to give the crawler permissions on AWS. Type the name `AWSGlueServiceRole-titanic`. Click **Next**.

5. On the next page, click on **Add database**. A new window will pop up. Type `bdok-database`, click **Create database**, and then close this window and go back to the **Glue crawler configuration** tab.

6. Back to the crawler, hit the refresh button and select your new **bdok-database** database. Leave the other options as the default. Click **Next**.

7. Now, in the last section, carefully review all the information and click **Create crawler**.

8. When it is ready, you will be taken to the crawler page on the AWS console. Click **Run crawler** to start the crawler. It should run for about 1 to 2 minutes (*Figure 9.5*).

Figure 9.5 – bdok-titanic-crawler

9. After the crawler is finished, you can validate that the table was correctly cataloged by accessing the **Data Catalog tables** menu item. The **titanic** table should be listed with the **bdok-database** database (*Figure 9.6*).

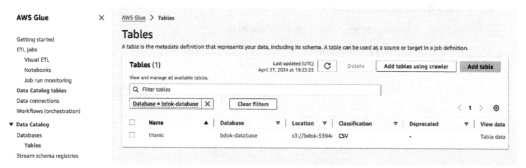

Figure 9.6 – Glue Data Catalog tables

10. Click on the `titanic` table's name to check whether the columns were correctly mapped.

11. Now, we need to create an IAM policy that gives Kubernetes permission to access the catalog and the data stored in S3. To that, in the console, go to the **IAM** page and select **Roles**. In the search box, type `studycluster` and you will see two roles created for Kubernetes, one service role and one node instance role. We want to change permissions on the node instance role (*Figure 9.7*).

Figure 9.7 – IAM Roles page

12. Click on the node instance role, then click on the **Add permissions** button and select **Create inline policy**.

13. On the **Specify permissions** page, click to edit as a JSON document and paste the JSON file available in this GitHub repository: `https://github.com/PacktPublishing/Bigdata-on-Kubernetes/blob/main/Chapter09/trino/iam/AthenaFullWithAllBucketsPolicy.json` (*Figure 9.8*). This policy allows Athena and Glue permissions, as well as getting all S3 data from any bucket. Remember that this is a very open policy and should not be used in production. It is a best security practice to allow access only in the needed buckets.

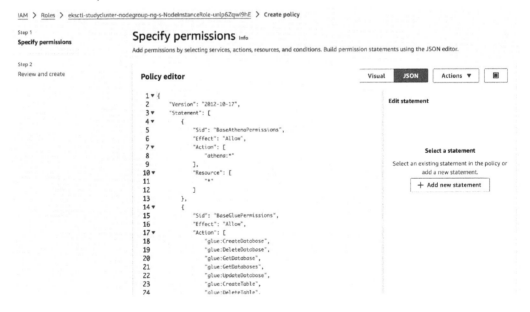

Figure 9.8 – Specifying permissions

14. Click **Next** and then save the policy as `AthenaFullWithAllBucketsPolicy` to facilitate this policy search later. Then, click on **Create policy**. And we are set to go!

Now, let's get back to DBeaver and play with some queries. First, we need to find where the table is stored. Expand the Trino connection in DBeaver and you will see a database named **hive**. This is where data from the Glue Data Catalog is mirrored in Trino. Expand **hive** and you will see the **bdok-database** catalog there. If you expand **Tables**, you will see the **titanic** dataset mapped.

To test a query, right-click the **hive** database and select **SQL Editor** and then **New SQL Script**. Now, run the query:

```
select * from hive."bdok-database".titanic
```

You should see the results (*Figure 9.9*):

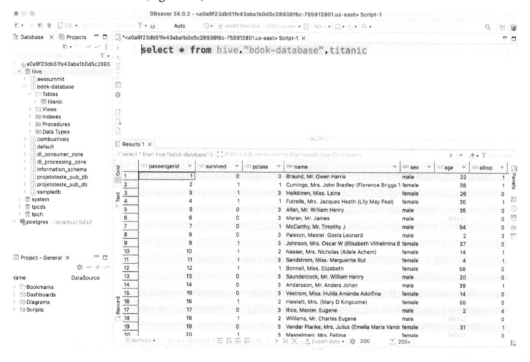

Figure 9.9 – DBeaver results from Trino

And, of course, Trino can perform any calculations or aggregations we like. Let's try a simple query to get the count and average age of all the passengers by `pclass` and `sex`. We will show the results ordered by `sex` and `pclass`.

```
select
    pclass,
    sex,
```

```
    COUNT(1) as people_count,
    AVG(age) as avg_age
from hive."bdok-database".titanic
group by pclass, sex
order by sex, pclass
```

This query yields this result:

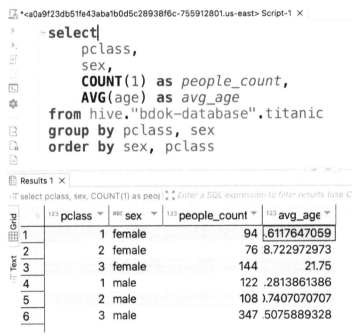

Figure 9.10 – Simple query on Titanic dataset

Now, let's visit Trino's monitoring page once again to see the query we just ran. Check the **Finished** box under **Query details** to see all queries; the first one shown is the query we just ran. Click on it to see the details.

That's it! You have successfully deployed Trino on Kubernetes and used it to query data from a data lake on Amazon S3. In the next section, we will work with Elasticsearch.

Deploying Elasticsearch in Kubernetes

While Trino provides a powerful SQL interface for querying structured data in your data lake, many modern applications also need to analyze semi-structured and unstructured data, such as logs, metrics, and text, in real time. For these types of use cases, Elasticsearch (or the ELK Stack, as it came to be known, referring to Elasticsearch, Logstash, and Kibana) provides a powerful solution.

Elasticsearch is an open source, distributed, RESTful search and analytics engine built on top of Apache Lucene. It is designed to store, search, and analyze large volumes of data quickly and in near real time.

At its core, Elasticsearch is a NoSQL database that uses JSON documents to represent data. It indexes all data in every field and uses advanced data structures and indexing techniques to make searches extremely fast.

How Elasticsearch stores, indexes and manages data

Data is stored in Elasticsearch in individual JSON documents. These documents are grouped into types within an index. You can think of an index like a database table that has a defined mapping or schema.

To add data to Elasticsearch, you make an HTTP request to the appropriate index with the JSON document in the request body. Elasticsearch automatically indexes all data in the document's fields using advanced data structures such as the inverted index from Apache Lucene.

This indexing process optimizes data for extremely fast queries and aggregations. Elasticsearch distributes data across shards, which can be allocated to different nodes in the cluster for redundancy and scalability.

When you want to query or retrieve data from Elasticsearch, you use the RESTful search API to define the query using a simple JSON request body. Results are returned in JSON format as well.

Elasticsearch is designed as a distributed system from the ground up. It can scale out to hundreds of servers and handle petabytes of data. Its core elements are as follows:

- **Nodes**, running instances of Elasticsearch that together form a **cluster**
- **Indexes**, which are collections of documents that have similar characteristics
- **Shards**, which are low-level partitions of an index that contain a slice of all the documents in the index
- **Replicas**, which are copies of a shard for redundancy and improved performance

At the core of Elasticsearch's distributed architecture is the sharding system. Sharding refers to horizontally splitting an Elasticsearch index into multiple pieces, called shards. This allows the index data to be distributed across multiple nodes in the cluster, providing several key benefits:

- **Horizontal scalability**: By distributing shards across nodes, Elasticsearch can effectively scale out to handle more data and higher query/indexing throughput. As the dataset grows, you can simply add more nodes to the cluster and Elasticsearch will automatically migrate shards to balance the load.

- **High availability**: Each shard can have one or more replica shards. A replica is a full copy of the primary shard. Replicas provide redundancy and high availability – if a node hosting a primary shard fails, Elasticsearch will automatically promote a replica as the new primary to take over.

- **Parallelization of operations**: Since index operations such as searches and aggregations are executed in parallel on each shard, having more shards allows greater parallelization and thus higher performance.

When you create an index in Elasticsearch, you need to specify the number of primary shards the index should have. For example, if you configure an index with three primary shards, Elasticsearch will horizontally partition the index data into three shards and distribute them across nodes in the cluster.

Each primary shard can also have zero or more replica shards configured. A common setup is to have one replica, meaning there are two copies of each shard – the primary and one replica. The replica shards are also distributed across nodes in the cluster, with each replica on a different node than its respective primary for redundancy.

Elasticsearch automatically manages shard allocation across nodes using a shard allocation strategy. The default is to spread shards across as many nodes as possible to balance the load. As nodes are added or removed from the cluster, Elasticsearch will automatically migrate shards to rebalance the cluster.

Queries are executed in parallel on each shard, with results being merged to produce the final result set. Writes (indexing new documents) are sent to a primary shard, which is responsible for validating the data, making changes persistent, and replicating changes to associated replica shards.

The number of shards configured for an index is fixed at index creation time. It cannot be changed later, so proper shard planning is important. Having more shards allows greater parallelization, but too many shards can also increase overhead.

A good rule of thumb is to start with enough shards (3 to 5 shards) that index data can be distributed across multiple nodes. The number can be increased if the index grows very large and more parallelization is needed. However, having hundreds or thousands of shards is generally not recommended due to increased cluster management overhead.

Now, let's see how to deploy Elasticsearch on Kubernetes.

Elasticsearch deployment

Here, we will work with **Elastic Cloud on Kubernetes** (**ECK**), an official Elastic operator that allows you to provision, manage, and orchestrate Elastic Stack applications on Kubernetes clusters. We will use the official Helm chart to install the operator. In your terminal, type the following:

```
helm repo add elastic https://helm.elastic.co
helm install elastic-operator elastic/eck-operator -n elastic
--create-namespace --version 2.12.1
```

This will download the Helm chart locally and deploy the default definitions for the Elastic Stack in a new environment named `elastic`. Here, we will use the `2.12.1` version of the Helm Chart.

Now, we will configure the deployment for an Elasticsearch cluster. The `elastic_cluster.yaml` YAML file does the trick.

```yaml
apiVersion: elasticsearch.k8s.elastic.co/v1
kind: Elasticsearch
metadata:
  name: elastic
spec:
  version: 8.13.0
  volumeClaimDeletePolicy: DeleteOnScaledownAndClusterDeletion
  nodeSets:
  - name: default
    count: 2
    podTemplate:
      spec:
        containers:
        - name: elasticsearch
          resources:
            requests:
              memory: 2Gi
              cpu: 1
            limits:
              memory: 2Gi
        initContainers:
        - name: sysctl
          securityContext:
            privileged: true
            runAsUser: 0
          command: ['sh', '-c', 'sysctl -w vm.max_map_count=262144']
    volumeClaimTemplates:
    - metadata:
        name: elasticsearch-data
      spec:
        accessModes:
        - ReadWriteOnce
        resources:
          requests:
            storage: 5Gi
        storageClassName: gp2
```

Let's take a closer look at this code. The first block specifies the API version and the kind of Kubernetes resource we are defining. In this case, it's an `Elasticsearch` resource from the `elasticsearch.k8s.elastic.co/v1` API group, which is provided by the ECK operator. The `metadata` block specifies the name of the cluster, in this case, `elastic`. Within the `spec` block, we set the Elasticsearch version to be used (`8.13.0`) and a policy that determines when the **PersistentVolumeClaims** (**PVCs**) associated with the Elasticsearch data volumes should be deleted. The `DeleteOnScaledownAndClusterDeletion` policy deletes the PVCs when the Elasticsearch cluster is scaled down or deleted entirely.

The `nodeSets` block defines the configuration for the Elasticsearch nodes. In this case, we have a single node set named `default` with a count of 2, meaning we will have two Elasticsearch nodes in the cluster. The `podTemplate` block specifies the configuration for the Pods that will run the Elasticsearch containers. Here, we define the resource requests and limits for the Elasticsearch container, setting the memory request and limit to 2 GiB and the CPU request to one vCPU.

The `initContainers` block is a recommendation from the official Elastic documentation for a production environment. It defines a container that will run before the main Elasticsearch container starts. In this case, we have an `initContainer` named `sysctl` that runs with privileged security context and sets the `vm.max_map_count` kernel setting to `262144`. This setting is recommended for running Elasticsearch on Linux to allow for a higher limit on memory-mapped areas in use.

Finally, the `volumeClaimTemplates` block defines the PVCs that will be used to store Elasticsearch data. In this case, we have a single PVC named `elasticsearch-data` with a requested storage size of 5 GiB. `accessModes` specifies that the volume should be `ReadWriteOnce`, meaning it can be mounted as read-write by a single node. `storageClassName` is set to `gp2`, which is an AWS EBS storage class for general-purpose SSD volumes.

After saving this file locally, run the following command to deploy an Elasticsearch cluster:

```
kubectl apply -f elastic_cluster.yaml -n elastic
```

Monitor the deployment with the following:

```
kubectl get pods -n elastic
```

Alternatively, you can use the following:

```
kubectl get elastic -n elastic
```

This will give a little more information. Note that this deployment will take a few minutes to finish. You can also get some detailed information on the cluster with the following:

```
kubectl describe elastic -n elastic
```

In the output, HEALTH should be green and the PHASE column should display Ready:

```
NAME        HEALTH    NODES    VERSION    PHASE
elastic     green     2        8.13.0     Ready
```

Now, let's move on to Kibana. We will follow the same process. The first thing to do is to set a YAML file named kibana.yaml with the deployment configuration.

```
apiVersion: kibana.k8s.elastic.co/v1
kind: Kibana
metadata:
  name: kibana
spec:
  version: 8.13.0
  count: 1
  elasticsearchRef:
    name: elastic
  http:
    service:
      spec:
        type: LoadBalancer
  podTemplate:
    spec:
      containers:
      - name: kibana
        env:
          - name: NODE_OPTIONS
            value: "--max-old-space-size=2048"
        resources:
          requests:
            memory: 1Gi
            cpu: 0.5
          limits:
            memory: 2Gi
            cpu: 2
```

This code is very similar to the previous one, but simpler. The main differences are in the spec block. First, the elasticsearchRef parameter specifies the name of the Elasticsearch cluster that Kibana should connect to. In this case, it's referencing the Elasticsearch cluster we created before named elastic. The http block configures the Kubernetes Service that will expose the Kibana deployment. Specifically, we are setting the type of the Service to LoadBalancer, which means that a load balancer will be provisioned by the cloud provider to distribute traffic across the Kibana instances. Finally, in the podTemplate block, we have an env configuration that sets an environment variable, NODE_OPTIONS, with the value --max-old-space-size=2048, which increases the maximum heap size for Kibana.

Now, we are ready to deploy:

```
kubectl apply -f kibana.yaml -n elastic
```

We use the same commands as before to monitor whether the deployment was successful. Now, we need to access the automatically generated password for Elastic and Kibana. We can do this with the following command:

```
kubectl get secret elastic-es-elastic-user -n elastic -o
go-template='{{.data.elastic | base64decode}}'
```

This command will print the generated password on the screen. Copy it and keep it safe. Now, run the following:

```
kubectl get svc -n elastic
```

To get the services list, copy the URL address for the LoadBalancer and paste it into a browser, adding :5601 at the end and https:// at the beginning. Kibana will not accept regular HTTP protocol connections. You should see the login page as in *Figure 9.11*.

Welcome to Elastic

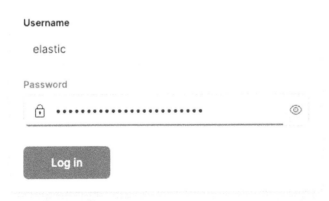

Figure 9.11 – Kibana login page

After inserting the username and password, you should be able to access Kibana's first empty page (*Figure 9.12*).

Welcome to Elastic

Start by adding integrations

Add data to your cluster from any source, then analyze and visualize it in real time. Use our solutions to add search anywhere, observe your ecosystem, and defend against security threats.

Add integrations   Explore on my own

Usage collection is enabled. This allows us to learn what our users are most interested in, so we can improve our products and services. Refer to our Privacy Statement ☑. Disable usage collection.

Figure 9.12 – Kibana's first empty page

Click on **Explore on my own** and you will now be able to play with Elastic as much as you want (although it does not have any data just yet). To do that, we will experiment with our (well-known) Titanic dataset. On the **Home** page, click on the menu in the upper-left corner and then click on **Stack Management** (the last option). On the next page, in the left menu, click on **Data Views** and then click on the **Upload a file** button in the center (*Figure 9.13*).

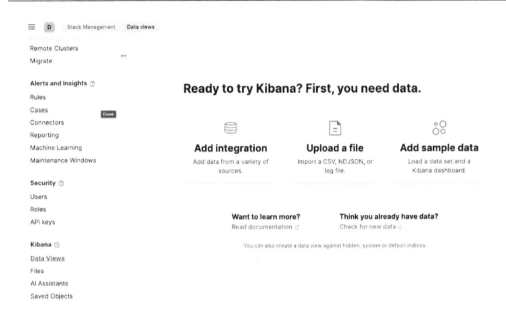

Figure 9.13 – Upload a file option in Kibana

Now, select the Titanic dataset CSV file you already have and upload it to Kibana. You will see a page with the mapped contents from the file (*Figure 9.14*).

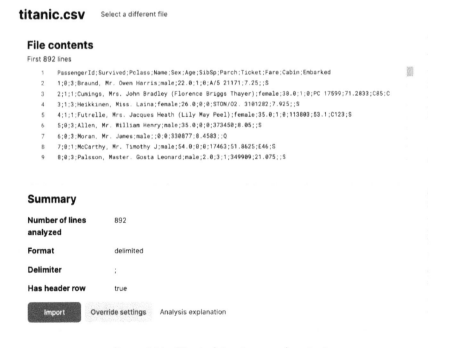

Figure 9.14 – Titanic dataset mapped contents

Now, click on **Import**. On the next page, you will be prompted for an index creation. Name the index `titanic` and make sure that the **Create data view** option is checked. Click on **Import** (*Figure 9.15*).

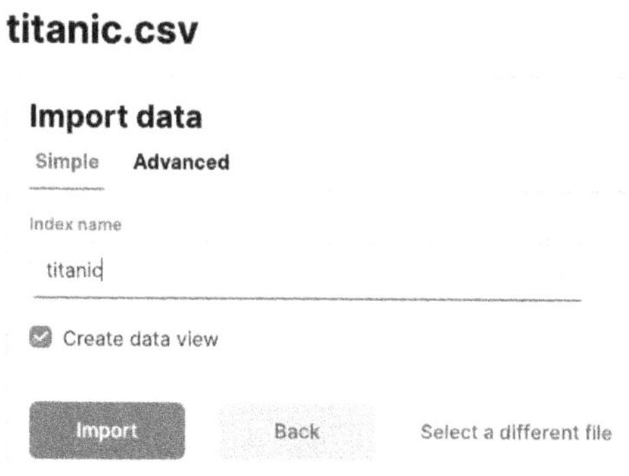

titanic.csv

Import data

Simple   **Advanced**

Index name

titanic

☑ Create data view

[Import]   Back   Select a different file

Figure 9.15 – Kibana – creating an index

In a few seconds, you should see a success screen. Now, let's play a little bit with this data with visualizations. Go back to the home page and, in the left menu, click on **Dashboards**. Then, click on **Create a dashboard** and then on **Create visualization**. This will get you to the visualization building in Kibana (*Figure 9.16*).

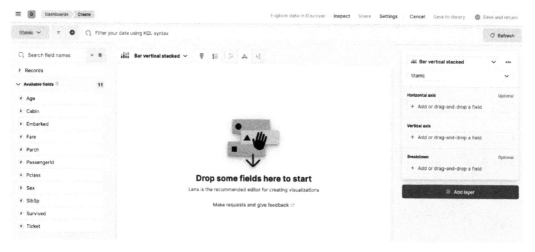

Figure 9.16 – Kibana visualization creation

Now, let's build some quick visualizations. On the right side of the page, select the type of visualization (let's keep **Bar vertical stacked**). For **Horizontal axis**, drag and drop the **Pclass** field. For **Vertical axis**, drag and drop the **Fare** field. As it is a numeric field, Kibana will automatically suggest the median as an aggregation function. Click on it to change it to **Average**. For **Breakdown**, drag and drop the **Sex** field. We should end up with a nice bar graph, as shown in *Figure 9.17*.

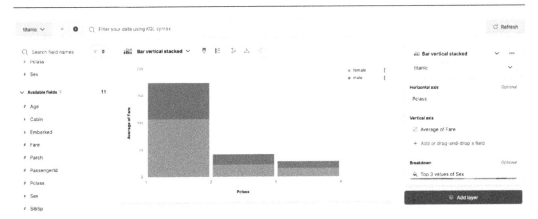

Figure 9.17 – Average fare price per sex and Pclass

Click on **Save and return** to view your newly created graphic on a new dashboard. Let's do another quick analysis. Click **Create visualization** again. This time, we will make a scatter plot with **Age** and **Fare** to see whether there is any correlation between those variables. Drop **Age** in **Horizontal axis** and **Fare** in **Vertical axis**. Click on **Vertical axis** to change the aggregation function to **Average**. Now, you have a nice scatter plot showing the interaction between those two variables. No significant correlation so far. Let's add the **Pclass** field as the breakdown and we will get a cool visualization of the data (*Figure 9.18*).

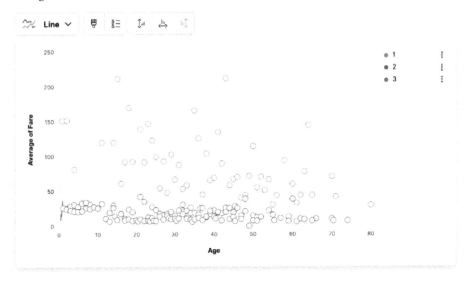

Figure 9.18 – Scatter plot in Kibana

Now, click on **Save and return** to see your new visualization on the dashboard. Finally, let's try something different. We will just show the number of Titanic survivors. Start a new visualization and, on the left menu, press the + button right next to the **Records** field. Kibana automatically suggests a simple **Legacy metric** visualization. In the top part, we will filter the field "Survived = 1" to count just the survivors. Click on **Update**. You should see the number **342** on the screen. Now, click on **Count of Records** in the right panel and change **Name** to `Survivors` (*Figure 9.19*).

Figure 9.19 – Survivors count visualization

Then, click **Save and return** and rearrange the dashboard manually as you wish (a simple example is shown in *Figure 9.20*).

Figure 9.20 – Your first Kibana dashboard

And that's it! You successfully deployed Elasticsearch and Kibana on Kubernetes, added data manually, and built a dashboard (with lots of potential). Feel free to play with Kibana, trying out other datasets and visualizations.

Summary

In this chapter, we explored two powerful tools, Trino and Elasticsearch, which enable effective data consumption and analysis in a Kubernetes-based big data architecture. We learned the importance of having a robust data consumption layer that bridges the gap between data repositories and business analysts, allowing them to extract valuable insights and make informed decisions.

We learned how to deploy Trino, a distributed SQL query engine, on Kubernetes and leverage its ability to directly query data stored in object storage systems such as Amazon S3. This eliminates the need for a traditional data warehouse and provides a cost-effective, scalable, and flexible solution for querying large datasets. We acquired hands-on experience in deploying Trino, configuring it to use the AWS Glue Data Catalog, and executing SQL queries against our data lake.

Additionally, we dove into Elasticsearch, a highly scalable and efficient search engine, along with Kibana, its powerful data visualization tool. We learned how to deploy Elasticsearch on Kubernetes using the ECK operator, index data for optimized storage and retrieval, and build simple yet insightful visualizations using Kibana. This combination equips us with the ability to analyze real-time data streams and uncover valuable patterns and trends.

The skills learned in this chapter are crucial in today's data-driven world, where organizations need to effectively consume and analyze vast amounts of data to make informed business decisions. Trino and Elasticsearch can also be extremely helpful for business teams who are not acquainted with coding to explore data (with simple SQL queries or in a visual way) and enhance their everyday decision-making.

In the next chapter, we will put all the pieces we have seen so far together to build a complete data pipeline on Kubernetes.

10

Building a Big Data Pipeline on Kubernetes

In the previous chapters, we covered the individual components required for building big data pipelines on Kubernetes. We explored tools such as Kafka, Spark, Airflow, Trino, and more. However, in the real world, these tools don't operate in isolation. They need to be integrated and orchestrated to form complete data pipelines that can handle various data processing requirements.

In this chapter, we will bring together all the knowledge and skills you have acquired so far and put them into practice by building two complete data pipelines: a batch processing pipeline and a real-time pipeline. By the end of this chapter, you will be able to (1) deploy and orchestrate all the necessary tools for building big data pipelines on Kubernetes; (2) write code for data processing, orchestration, and querying using Python, SQL, and APIs; (3) integrate different tools seamlessly to create complex data pipelines; (4) understand and apply best practices for building scalable, efficient, and maintainable data pipelines.

We will start by ensuring that all the required tools are deployed and running correctly in your Kubernetes cluster. Then, we will dive into building the batch processing pipeline, where you will learn how to ingest data from various sources, process it using Spark, and store the results in a data lake for querying and analysis.

Next, we will tackle the real-time pipeline, which is essential for processing and analyzing data streams in near real time. You will learn how to ingest and process data streams using Kafka, Spark Streaming, and Elasticsearch, enabling you to build applications that can react to events as they occur.

By the end of this chapter, you will have gained hands-on experience in building complete data pipelines on Kubernetes, preparing you for real-world big data challenges. Let's dive in and unlock the power of big data on Kubernetes!

In this chapter, we're going to cover the following main topics:

- Checking the deployed tools
- Building a batch pipeline
- Building a real-time pipeline

Technical requirements

For the activities in this chapter, you should have a running Kubernetes cluster. Refer to *Chapter 8* for details on Kubernetes deployment and all the necessary operators. You should also have an **Amazon Web Services** (**AWS**) account to run the exercises. We will also use DBeaver to check data. For installation instructions, please refer to *Chapter 9*.

All code for this chapter is available at `https://github.com/PacktPublishing/Bigdata-on-Kubernetes` in the `Chapter10` folder.

Checking the deployed tools

Before we get our hands into a fully orchestrated data pipeline, we need to make sure that all the necessary operators are correctly deployed on Kubernetes. We will check for the Spark operator, the Strimzi operator, Airflow, and Trino. First, we'll check for the Spark operator using the following command:

```
kubectl get pods -n spark-operator
```

This output shows that the Spark operator is successfully running:

```
NAME                                    READY   STATUS
spark-operator-74db6fcf98-f86vt         1/1     Running
spark-operator-webhook-init-5594s       0/1     Completed
```

Now, we will check Trino. For that, type the following:

```
kubectl get pods -n trino
```

Check if all pods are correctly running; in our case, one coordinator pod and two worker pods. Also, check for Kafka and Elasticsearch with the following commands:

```
kubectl get pods -n kafka
kubectl get pods -n elastic
```

Last, we will need a new deployment of Airflow. We will need to use a specific version of Airflow and one of its providers' libraries to work correctly with Spark. I have already set up an image of Airflow 2.8.1 with the 7.13.0 version of the `apache-airflow-providers-cncf-kubernetes` library (needed for `SparkKubernetesOperator`). If you have Airflow already installed, let's delete it with the following command:

```
helm delete airflow -n airflow
```

Make sure that all services and persistent volume claims are deleted as well, using the following code:

```
kubectl delete svc --all -n airflow
kubectl delete pvc --all -n airflow
```

Then, we need to change slightly the configuration we already have for the `custom_values.yaml` file. We need to set the `defaultAirflowTag` and the `airflowVersion` parameters to `2.8.1`, and we will change the `images.airflow` parameter to get an already prepared public image:

```
images:
  airflow:
    repository: "docker.io/neylsoncrepalde/apache-airflow"
    tag: "2.8.1-cncf7.13.0"
    digest: ~
    pullPolicy: IfNotPresent
```

Also, don't forget to adjust the `dags.gitSync` parameter if you are working with a different GitHub repo or folder. A complete version of the adapted `custom_values.yaml` code is available at `https://github.com/PacktPublishing/Bigdata-on-Kubernetes/tree/main/Chapter10/airflow_deployment`. Redeploy Airflow with the new configurations as follows:

```
helm install airflow apache-airflow/airflow --namespace airflow
--create-namespace -f custom_values.yaml
```

The last configurations needed allow Airflow to run `SparkApplication` instances on the cluster. We will set up a service account and a cluster role binding for running Spark on the Airflow namespace:

```
kubectl create serviceaccount spark -n airflow
kubectl create clusterrolebinding spark-role --clusterrole=edit
--serviceaccount=airflow:spark --namespace=airflow
```

Now, we will create a new cluster role and a cluster role binding to give Airflow workers the necessary permissions. Set up a YAML file:

rolebinding_for_airflow.yaml

```
apiVersion: rbac.authorization.k8s.io/v1
kind: ClusterRole
metadata:
  name: spark-cluster-cr
  labels:
```

```
        rbac.authorization.kubeflow.org/aggregate-to-kubeflow-edit: "true"
rules:
  - apiGroups:
      - sparkoperator.k8s.io
    resources:
      - sparkapplications
    verbs:
      - "*"
---
apiVersion: rbac.authorization.k8s.io/v1
kind: ClusterRoleBinding
metadata:
  name: airflow-spark-crb
roleRef:
  apiGroup: rbac.authorization.k8s.io
  kind: ClusterRole
  name: spark-cluster-cr
subjects:
  - kind: ServiceAccount
    name: airflow-worker
    namespace: airflow
```

Now, deploy this configuration with the following:

```
kubectl apply -f rolebinding_for_airflow.yaml -n airflow
```

That's it! We can now move to the implementation of a batch data pipeline. Let's get to it.

Building a batch pipeline

For the batch pipeline, we will use the IMBD dataset we worked on in *Chapter 5*. We are going to automate the whole process from data acquisition and ingestion into our data lake on **Amazon Simple Storage Service** (**Amazon S3**) up to the delivery of consumption-ready tables in Trino. In *Figure 10.1*, you can see a diagram representing the architecture for this section's exercise:

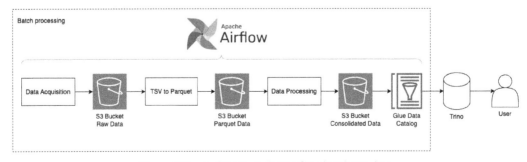

Figure 10.1 – Architecture design for a batch pipeline

Now, let's get to the code.

Building the Airflow DAG

Let's start developing our Airflow DAG as usual. The complete code is available at `https://github.com/PacktPublishing/Bigdata-on-Kubernetes/tree/main/Chapter10/batch/dags` folder:

1. The first lines of the Airflow DAG are shown next:

    imdb_dag.py

    ```
    from airflow.decorators import task, dag
    from airflow.utils.task_group import TaskGroup
    from airflow.providers.cncf.kubernetes.operators.spark_
    kubernetes import SparkKubernetesOperator
    from airflow.providers.cncf.kubernetes.sensors.spark_kubernetes
    import SparkKubernetesSensor
    from airflow.providers.amazon.aws.operators.glue_crawler import
    GlueCrawlerOperator
    from airflow.models import Variable
    from datetime import datetime
    import requests
    import boto3


    aws_access_key_id = Variable.get("aws_access_key_id")
    aws_secret_access_key = Variable.get("aws_secret_access_key")


    s3 = boto3.client('s3',
        aws_access_key_id=aws_access_key_id,
        aws_secret_access_key=aws_secret_access_key
    )


    default_args = {
        'owner': 'Ney',
        'start_date': datetime(2024, 5, 10)
    }
    ```

    This code imports the necessary libraries, sets up two environment variables needed to authenticate on AWS, defines an Amazon S3 client, and sets some default configurations.

2. In the next block, we will start the DAG function in the code:

    ```
    @dag(
            default_args=default_args,
            schedule_interval="@once",
            description="IMDB Dag",
    ```

```
        catchup=False,
        tags=['IMDB']
)
def IMDB_batch():
```

This block integrates default arguments for the DAG and defines a schedule interval to run only once and some metadata.

3. Now, we will define the first task that will automatically download the datasets and store them on S3 (the first line is repeated):

```
def IMDB_batch():

    @task
    def data_acquisition():
        urls_dict = {
            "names.tsv.gz": "https://datasets.imdbws.com/
name.basics.tsv.gz",
            "basics.tsv.gz": "https://datasets.imdbws.com/
title.basics.tsv.gz",
            "crew.tsv.gz": "https://datasets.imdbws.com/
title.crew.tsv.gz",
            "principals.tsv.gz": "https://datasets.imdbws.
com/title.principals.tsv.gz",
            "ratings.tsv.gz": "https://datasets.imdbws.com/
title.ratings.tsv.gz"
        }

        for title, url in urls_dict.items():
            response = requests.get(url, stream=True)
            with open(f"/tmp/{title}", mode="wb") as file:
                file.write(response.content)
            s3.upload_file(f"/tmp/{title}", "bdok-<YOUR_ACCOUNT_
NUMBER>", f"landing/imdb/{title}")

        return True
```

This code is derived from what we developed in *Chapter 5*, with a small modification at the end to upload the downloaded files to S3.

4. Next, we will call Spark processing jobs to transform that data. The first step is only read the data in its original format (TSV) and transform it to Parquet (which is optimized for storage and processing in Spark). First, we define a `TaskGroup` instance to better organize the tasks:

```
with TaskGroup("tsvs_to_parquet") as tsv_parquet:
    tsvs_to_parquet = SparkKubernetesOperator(
        task_id="tsvs_to_parquet",
```

```
        namespace="airflow",
        #application_file=open(f"{APP_FILES_PATH}/spark_imdb_
tsv_parquet.yaml").read(),
        application_file="spark_imdb_tsv_parquet.yaml",
        kubernetes_conn_id="kubernetes_default",
        do_xcom_push=True
    )
    tsvs_to_parquet_sensor = SparkKubernetesSensor(
        task_id="tsvs_to_parquet_sensor",
        namespace="airflow",
        application_name="{{ task_instance.xcom_pull(task_
ids='tsvs_to_parquet.tsvs_to_parquet')['metadata']['name'] }}",
        kubernetes_conn_id="kubernetes_default"
    )
    tsvs_to_parquet >> tsvs_to_parquet_sensor
```

Within this group, there are two tasks:

- `tsvs_to_parquet`: This is a `SparkKubernetesOperator` task that runs a Spark job on Kubernetes. The job is defined in the `spark_imdb_tsv_parquet.yaml` file, which contains the Spark application configuration. We use the `do_xcom_push=True` parameter, which enables cross-communication between this and the following task.

- `tsvs_to_parquet_sensor`: This is a `SparkKubernetesSensor` task that monitors the Spark job launched by the `tsvs_to_parquet` task. It retrieves the Spark application name from the metadata pushed by the previous task using the `task_instance.xcom_pull` method. This sensor waits for the Spark job to complete before allowing the DAG to proceed to the next tasks.

The `tsvs_to_parquet >> tsvs_to_parquet_sensor` line sets up the task dependency, ensuring that the `tsvs_to_parquet_sensor` task runs after the `tsvs_to_parquet` task completes successfully.

5. Next, we have another round of data processing with Spark. This time, we will join all the tables to build a consolidated unique table. This consolidated form has been called **One Big Table** (**OBT**) in the market. For that, we will define a new `TaskGroup` instance called `Transformations` and proceed the same as the previous code block:

```
with TaskGroup('Transformations') as transformations:
    consolidated_table = SparkKubernetesOperator(
        task_id='consolidated_table',
        namespace="airflow",
        application_file="spark_imdb_consolidated_table.yaml",
        kubernetes_conn_id="kubernetes_default",
        do_xcom_push=True
    )
```

```
consolidated_table_sensor = SparkKubernetesSensor(
    task_id='consolidated_table_sensor',
    namespace="airflow",
    application_name="{{ task_instance.xcom_pull(task_
ids='Transformations.consolidated_table')['metadata']['name']
}}",
    kubernetes_conn_id="kubernetes_default"
)
consolidated_table >> consolidated_table_sensor
```

6. Finally, after the data is processed and written in Amazon S3, we will trigger a Glue crawler that will write the metadata for this table into Glue Data Catalog, making it available for Trino:

```
glue_crawler_consolidated = GlueCrawlerOperator(
    task_id='glue_crawler_consolidated',
    region_name='us-east-1',
    aws_conn_id='aws_conn',
    wait_for_completion=True,
    config = {'Name': 'imdb_consolidated_crawler'}
)
```

Remember that all of this code should be indented to be inside the IMDB_Batch() function.

7. Now, in the last block of this code, we will configure the dependencies between tasks and TaskGroup instances and trigger the execution of the DAG function:

```
da = data_acquisition()
da >> tsv_parquet >> transformations
transformations >> glue_crawler_consolidated

execution = IMDB_batch()
```

Now, we have to set up the two SparkApplication instances Airflow is going to call and the Glue crawler on AWS. Let's get to it.

Creating SparkApplication jobs

We will follow the same pattern used in *Chapter 8* to configure the Spark jobs. We need PySpark code that will be stored in S3 and a YAML file for the job definition that must be in the dags folder, along with the Airflow DAG code:

1. As the YAML file is very similar to what we did before, we will not get into details here. The code for both YAML files is available at https://github.com/PacktPublishing/ Bigdata-on-Kubernetes/tree/main/Chapter10/batch/dags folder. Create those files and save them as spark_imdb_tsv_parquet.yaml and spark_imdb_ consolidated_table.yaml in the dags folder.

2. Next, we will take a look at the PySpark code. The first job is quite simple. It reads the data from the TSV files ingested by Airflow and writes back the same data transformed to Parquet. First, we import Spark modules and define a `SparkConf` object with the necessary configurations for the Spark application:

```python
from pyspark import SparkContext, SparkConf
from pyspark.sql import SparkSession

conf = (
    SparkConf()
        .set("spark.cores.max", "2")
        .set("spark.executor.extraJavaOptions", "-Dcom.
amazonaws.services.s3.enableV4=true")
        .set("spark.driver.extraJavaOptions", "-Dcom.amazonaws.
services.s3.enableV4=true")
        .set("spark.hadoop.fs.s3a.fast.upload", True)
        .set("spark.hadoop.fs.s3a.impl", "org.apache.hadoop.
fs.s3a.S3AFileSystem")
        .set("spark.hadoop.fs.s3a.aws.credentials.provider",
"com.amazonaws.auth.EnvironmentVariablesCredentials")
        .set("spark.jars.packages", "org.apache.hadoop:hadoop-
aws:2.7.3")
)


sc = SparkContext(conf=conf).getOrCreate()
```

These configurations are specific to working with Amazon S3 and enabling certain features such as S3 V4 authentication, fast uploads, and using the S3A filesystem implementation. The `spark.cores.max` property limits the maximum number of cores used by the application to 2. The last line creates a `SparkContext` object with the configurations defined before.

3. Next, we create a `SparkSession` instance and set the log level to `"WARN"` so that only warning and error messages get displayed in the logs. This is good for log readability:

```python
if __name__ == "__main__":
    spark = SparkSession.builder.appName("SparkApplicationJob").
getOrCreate()
    spark.sparkContext.setLogLevel("WARN")
```

4. Next, we will define table schemas. This is extremely important when working with large datasets as it improves Spark's performance when dealing with text-based files (such as TSV, CSV, and so on). Next, we present only the schema for the first table to simplify readability. The full code can be found at `https://github.com/PacktPublishing/Bigdata-on-Kubernetes/tree/main/Chapter10/batch/spark_code` folder:

```python
schema_names = "nconst string, primaryName string, birthYear
int, deathYear int, primaryProfession string, knownForTitles
string"
```

5. Now, we read the table into a Spark DataFrame (also displaying only the reading of the first table):

```
names = (
    spark
    .read
    .schema(schema_names)
    .options(header=True, delimiter="\t")
    .csv('s3a://bdok-<ACCOUNT_NUMBER>/landing/imdb/names.tsv.
gz')
)
```

6. Next, we write the tables back to S3 in Parquet:

```
names.write.mode("overwrite").parquet("s3a://bdok-<ACCOUNT_
NUMBER>/bronze/imdb/names")
```

7. Finally, we stop the Spark session and release any resources used by the application:

```
spark.stop()
```

8. Save this file as spark_imdb_tsv_parquet.py and upload it to the S3 bucket you defined in the YAML file (in this case, s3a://bdok-<ACCOUNT_NUMBER>/spark-jobs/).

9. Now, we will define the second SparkApplication instance responsible for building the OBT. For this second code, we will skip the Spark configuration and SparkSession code blocks as they are almost the same as the last job except for one import we must do:

```
from pyspark.sql import functions as f
```

This imports the functions module that will allow data transformations using Spark internals.

10. We begin here by reading the datasets:

```
names = spark.read.parquet("s3a://bdok-<ACCOUNT_NUMBER>/bronze/
imdb/names")
basics = spark.read.parquet("s3a://bdok-<ACCOUNT_NUMBER>/bronze/
imdb/basics")
crew = spark.read.parquet("s3a://bdok-<ACCOUNT_NUMBER>/bronze/
imdb/crew")
principals = spark.read.parquet("s3a://bdok-<ACCOUNT_NUMBER>/
bronze/imdb/principals")
ratings = spark.read.parquet("s3a://bdok-<ACCOUNT_NUMBER>/
bronze/imdb/ratings")
```

11. The knownForTitles column in the names dataset and the directors column in the crew dataset have several values in the same cell that need to be exploded to get one line per director and titles:

```
names = names.select(
    'nconst', 'primaryName', 'birthYear', 'deathYear',
```

```
      f.explode(f.split('knownForTitles', ',')).
alias('knownForTitles')
)

crew = crew.select(
    'tconst', f.explode(f.split('directors', ',')).
alias('directors'), 'writers'
)
```

12. Now, we begin to join tables:

```
basics_ratings = basics.join(ratings, on=['tconst'],
how='inner')
principals_names = (
    principals.join(names, on=['nconst'], how='inner')
    .select('nconst', 'tconst','ordering', 'category',
'characters', 'primaryName', 'birthYear', 'deathYear')
    .dropDuplicates()
)
directors = (
    crew
    .join(names, on=crew.directors == names.nconst, how='inner')
    .selectExpr('tconst', 'directors', 'primaryName as
directorPrimaryName',
                'birthYear as directorBirthYear', 'deathYear as
directorDeathYear')
    .dropDuplicates()
)
```

Here, we perform three join operations: (a) basics_ratings is created by joining the basics and ratings DataFrames on the tconst column (a movie identifier); (b) principals_names is created by joining the principals and names DataFrames on the nconst column (an actor identifier); we select some specific columns and remove duplicates; (c) a directors table is created by joining the crew and names DataFrames, where the directors column in crew matches the nconst column in names. Then, we select specific columns, rename some columns so that we can identify which data relates specifically to directors, and remove duplicates.

13. Next, we will create a basics_principals table joining the crew and principals_names datasets to get a complete dataset on crew and movie performers. Finally, we create a basics_principals_directors table joining the directors table information:

```
basics_principals = basics_ratings.join(principals_names,
on=['tconst'], how='inner').dropDuplicates()
basics_principals_directors = basics_principals.join(directors,
on=['tconst'], how='inner').dropDuplicates()
```

14. Finally, we write this final table as a Parquet file on Amazon S3 and stop the Spark job:

```
basics_principals_directors.write.mode("overwrite").
parquet("s3a://bdok-<ACCOUNT_NUMBER>/silver/imdb/consolidated")
spark.stop()
```

The last thing to do is to create a Glue crawler that will make the information on the OBT available for Trino.

Creating a Glue crawler

We will create a Glue crawler using the AWS console. Follow the next steps:

1. Log in to AWS and go to the **AWS Glue** page. Then, click on **Crawlers** in the side menu and click on **Create crawler** (*Figure 10.2*):

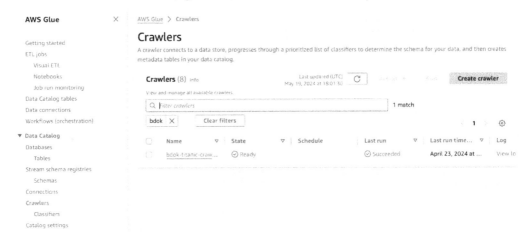

Figure 10.2 – AWS Glue: Crawlers page

2. Next, type `imdb_consolidated_crawler` (same name as referenced in the Airflow code) for the crawler's name and a description as you like. Click **Next**.

3. Make sure **Not yet** is checked for the first configuration, **Is your data already mapped to Glue tables?**. Then, click on **Add a data source** (*Figure 10.3*):

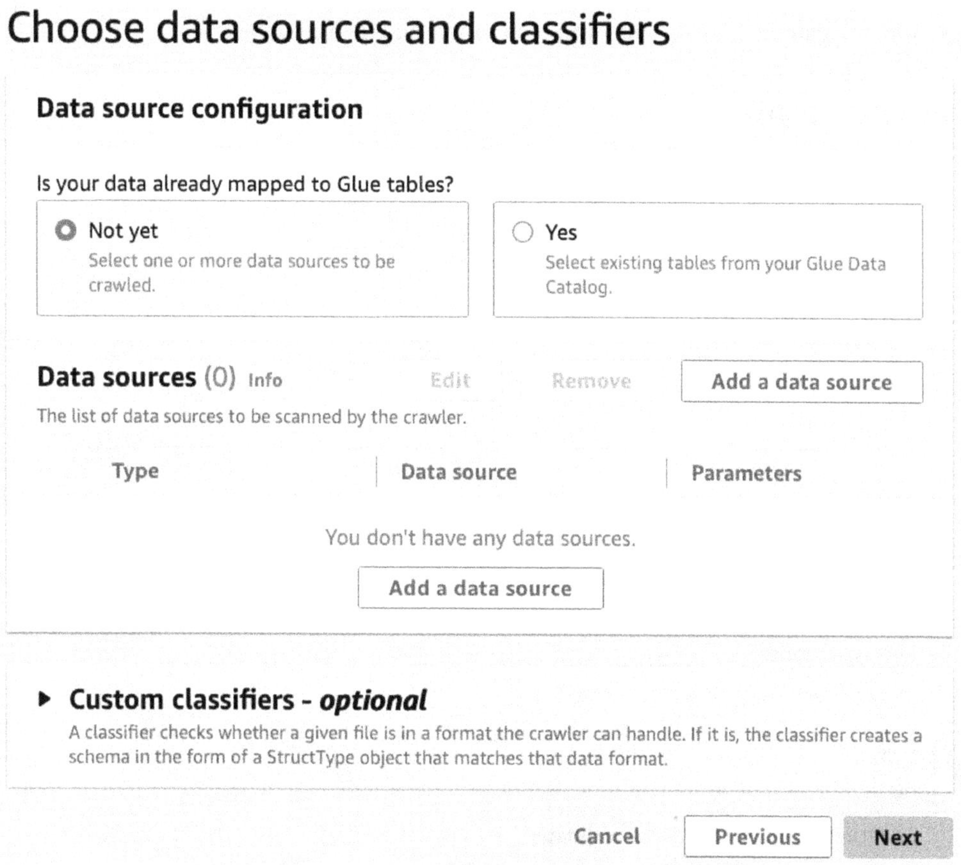

Choose data sources and classifiers

Data source configuration

Is your data already mapped to Glue tables?

○ **Not yet**
Select one or more data sources to be crawled.

○ Yes
Select existing tables from your Glue Data Catalog.

Data sources (0) Info   Edit   Remove   **Add a data source**

The list of data sources to be scanned by the crawler.

Type	Data source	Parameters

You don't have any data sources.

Add a data source

▶ **Custom classifiers - *optional***

A classifier checks whether a given file is in a format the crawler can handle. If it is, the classifier creates a schema in the form of a StructType object that matches that data format.

Cancel   Previous   **Next**

Figure 10.3 – Adding a data source

4.   In the **Add data source** pop-up page, make sure **S3** is selected. Leave **Network connection** blank, and in the **S3 path** field, type the URL to the S3 bucket we are going to write the IMDB OBT to (`s3://bdok-<ACCOUNT_NUMBER>/silver/imdb/consolidated`), as shown in *Figure 10.4*. Click **Add an S3 data source** and then click **Next**:

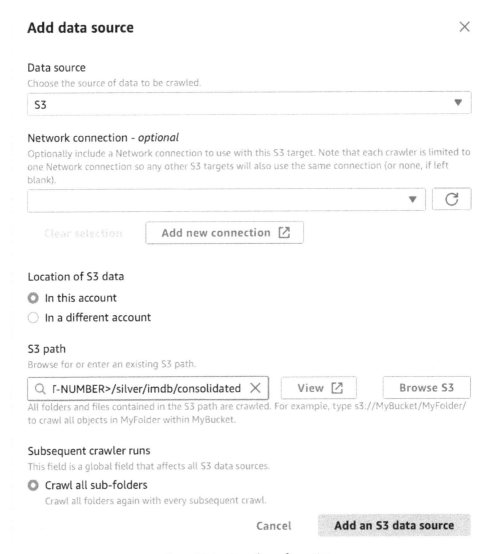

Figure 10.4 – S3 path configuration

5. On the next page, click on **Create new IAM role** and fill it with the name you like. Be sure it is not an existing role name. Click **Next** (*Figure 10.5*):

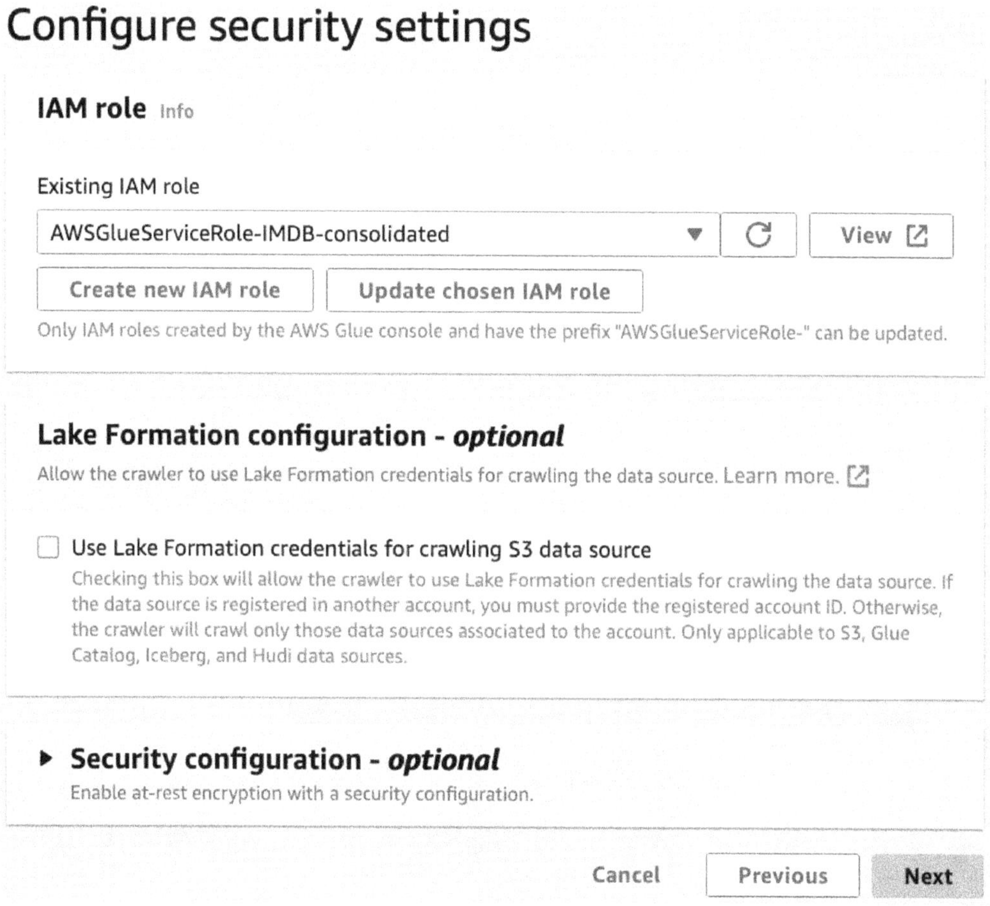

Configure security settings

IAM role Info

Existing IAM role

| AWSGlueServiceRole-IMDB-consolidated ▼ | C | View ☑ |

| Create new IAM role | Update chosen IAM role |

Only IAM roles created by the AWS Glue console and have the prefix "AWSGlueServiceRole-" can be updated.

Lake Formation configuration - *optional*

Allow the crawler to use Lake Formation credentials for crawling the data source. Learn more. ☑

☐ Use Lake Formation credentials for crawling S3 data source

  Checking this box will allow the crawler to use Lake Formation credentials for crawling the data source. If the data source is registered in another account, you must provide the registered account ID. Otherwise, the crawler will crawl only those data sources associated to the account. Only applicable to S3, Glue Catalog, Iceberg, and Hudi data sources.

▶ **Security configuration - *optional***
  Enable at-rest encryption with a security configuration.

| Cancel | Previous | **Next** |

Figure 10.5 – IAM role configuration

6. On the next page, you can choose the same database we created in *Chapter 9* to work with Trino (bdok-database). For **Table name prefix**, I suggest putting imdb- to make it easier to locate this table (*Figure 10.6*). Leave the **Crawler schedule** setting as **On demand**. Click **Next**:

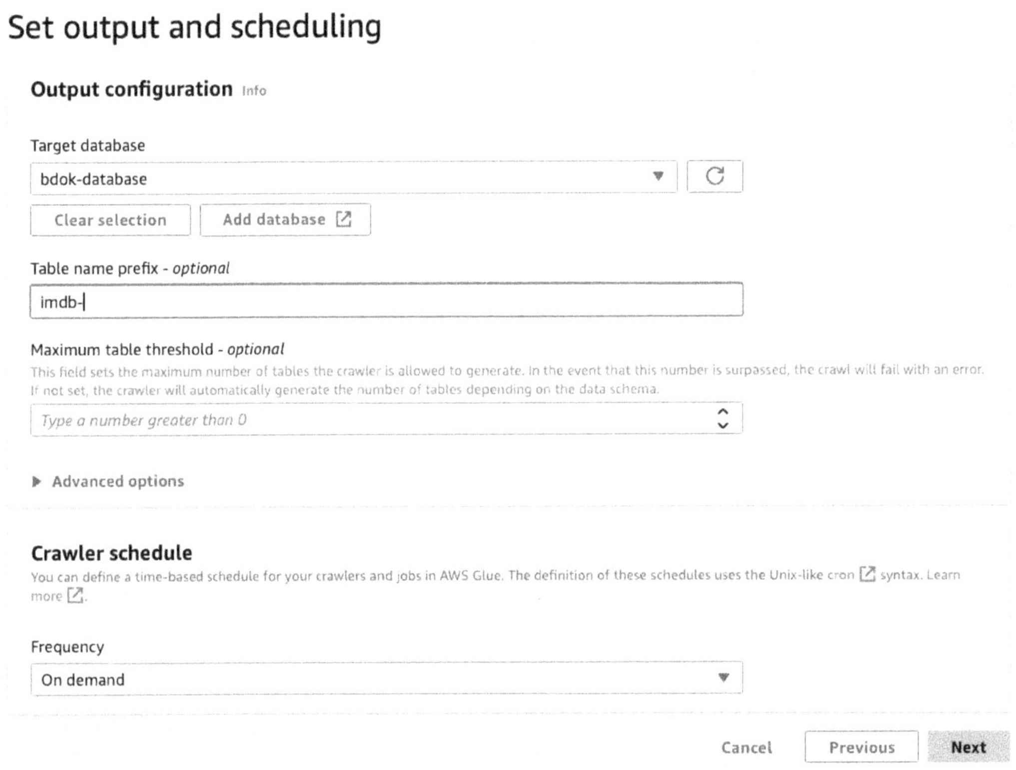

Set output and scheduling
Output configuration Info
Target database
bdok-database
Clear selection
Add database
Table name prefix - optional
imdb-
Maximum table threshold - optional
This field sets the maximum number of tables the crawler is allowed to generate. In the event that this number is surpassed, the crawl will fail with an error. If not set, the crawler will automatically generate the number of tables depending on the data schema.
Type a number greater than 0
Advanced options
Crawler schedule
You can define a time-based schedule for your crawlers and jobs in AWS Glue. The definition of these schedules uses the Unix-like cron syntax. Learn more.
Frequency
On demand
Cancel
Previous
Next

Figure 10.6 – Target database and table name prefix

7. In the final step, review all the information provided. If all is correct, click **Create crawler**.

That's it! All set. Now, we go back to the Airflow UI in a browser and activate the DAG to see the "magic" happening (*Figure 10.7*):

IMDB_batch
2024-05-09, 21:00:00 -03
SparkKubernetesOperator
Details
Graph
Gantt
Code
tsvs_to_parquet
success
tsvs_to-parquet_sensor
Layout:
Left -> Right
data_acqusition
success
@task
tsvs_to_parquet
success
- 2 tasks
tsvs_to_parquet
success
SparkKubernetesoperator
tsvs_to-parquet_sensor
success
SparkKubernetesSensor
Transformations
success
- 2 tasks
consolidated_table
success
SparkKubernetesoperator
consolidated_table_sensor
success
SparkKubernetesSensor
glue_crawler_consolidated
success
GlueCrawlerOperator

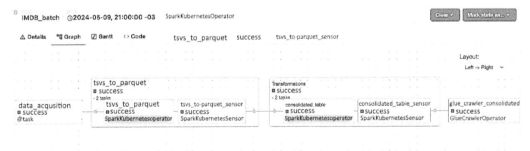

Figure 10.7 – Running the complete DAG on Airflow

After the DAG is successful, wait about 2 minutes for the crawler to stop, and then let's search for our data using DBeaver. Let's play a little bit and search for all John Wick movies (*Figure 10.8*):

Figure 10.8 – Checking the OBT in Trino with DBeaver

Et voilà! You have just run your complete batch data processing pipeline connecting all the batch tools we studied so far. Congratulations! Now, we will move to building a data streaming pipeline using Kafka, Spark Streaming, and Elasticsearch.

Building a real-time pipeline

For the real-time pipeline, we will use the same data simulation code we used in *Chapter 8* with an enhanced architecture. In *Figure 10.9*, you can find an architecture design of the pipeline we are about to build:

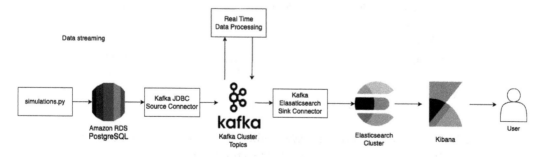

Figure 10.9 – Real-time data pipeline architecture

First thing, we need to create a **virtual private cloud** (**VPC**) – a private network – on AWS and set up a **Relational Database Service** (**RDS**) Postgres database that will work as our data source:

1. Go to the AWS console and navigate to the **VPC** page. On the **VPC** page, click on **Create VPC**, and you will get to the configuration page.

2. Make sure **VPC and more** is selected. Type bdok in the **Name tag auto-generation** block and check the **Auto-generate** box so that AWS will generate all resources' names according to the project name. For **IPv4 CIDR block**, let's use the 10.20.0.0/16 **Classless Inter-Domain Routing** (**CIDR**) block. Leave the rest as default (*Figure 10.10*):

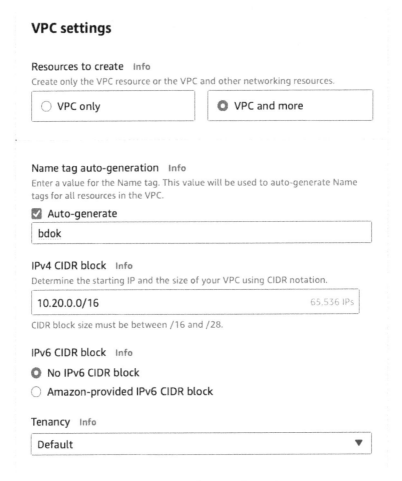

Figure 10.10 – VPC basic configurations

3. Now, roll the page. You can leave the **Availability Zones** (**AZs**) and subnets configuration as they are (two AZs, two public subnets, and two private subnets). Make sure to mark **In 1**

AZ for the **network address translation** (**NAT**) gateway. Leave the **S3 Gateway** box marked (*Figure 10.11*). Also, leave the two DNS options marked at the end. Click on **Create VPC**:

Figure 10.11 – NAT gateway configuration

4. The VPC will take a few minutes to create. After it is successfully created, in the AWS console, navigate to the **RDS** page, click on **Databases** in the side menu, and then click on **Create Database**.

5. On the next page, choose **Standard create** and choose **Postgres** for the database. Leave the default engine version. In the **Templates** section, choose **Free tier** because we only need a small database for this exercise.

6. In the **Settings** section, choose a name for our database. In this case, I'm working with bdok-postgres. For the credentials, leave postgres as the master username, check **Self managed** for the **Credentials management** option, and choose a master password (*Figure 10.12*):

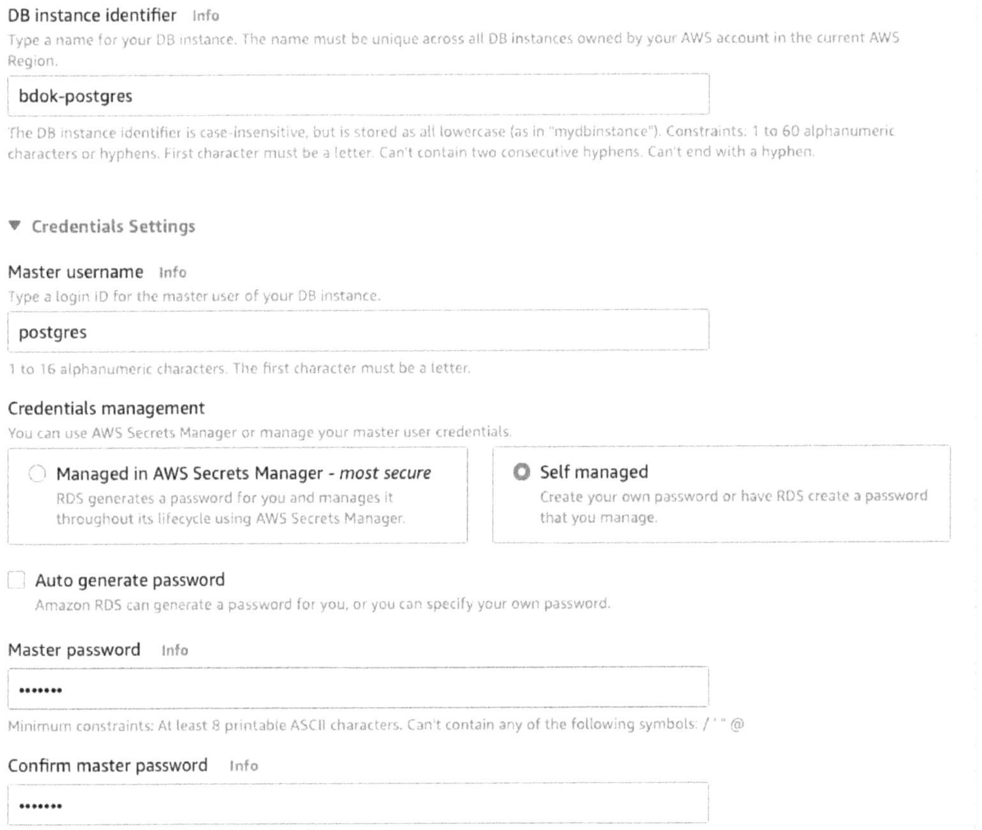

DB instance identifier   Info
Type a name for your DB instance. The name must be unique across all DB instances owned by your AWS account in the current AWS Region.

bdok-postgres

The DB instance identifier is case-insensitive, but is stored as all lowercase (as in "mydbinstance"). Constraints: 1 to 60 alphanumeric characters or hyphens. First character must be a letter. Can't contain two consecutive hyphens. Can't end with a hyphen.

▼ Credentials Settings

Master username   Info
Type a login ID for the master user of your DB instance.

postgres

1 to 16 alphanumeric characters. The first character must be a letter.

Credentials management
You can use AWS Secrets Manager or manage your master user credentials.

◯ Managed in AWS Secrets Manager - *most secure*
    RDS generates a password for you and manages it throughout its lifecycle using AWS Secrets Manager.

🔘 Self managed
    Create your own password or have RDS create a password that you manage.

☐ Auto generate password
    Amazon RDS can generate a password for you, or you can specify your own password.

Master password   Info

•••••••

Minimum constraints: At least 8 printable ASCII characters. Can't contain any of the following symbols: / ' " @

Confirm master password   Info

•••••••

Figure 10.12 – Database name and credentials

7. Leave the **Instance configuration** and **Storage** sections as default.

8. In the **Connectivity** section, choose **Don't connect to an EC2 compute resource** as this won't be needed for our exercise. On the **VPC** page, choose the VPC we just created (bdok-vpc) and leave the **Create new DB Subnet Group** option as default. In the **VPC security group** section, choose **Create new** and type bdok-database-sg for the security group name (*Figure 10.13*):

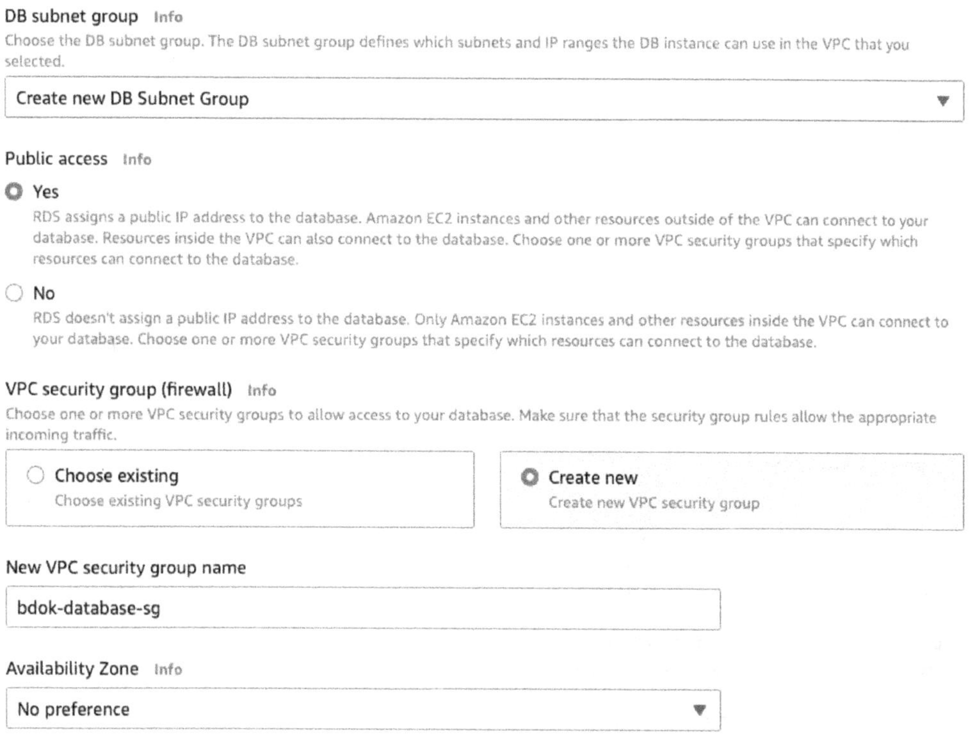

DB subnet group Info

Choose the DB subnet group. The DB subnet group defines which subnets and IP ranges the DB instance can use in the VPC that you selected.

Create new DB Subnet Group   ▼

Public access Info

● Yes

RDS assigns a public IP address to the database. Amazon EC2 instances and other resources outside of the VPC can connect to your database. Resources inside the VPC can also connect to the database. Choose one or more VPC security groups that specify which resources can connect to the database.

○ No

RDS doesn't assign a public IP address to the database. Only Amazon EC2 instances and other resources inside the VPC can connect to your database. Choose one or more VPC security groups that specify which resources can connect to the database.

VPC security group (firewall) Info

Choose one or more VPC security groups to allow access to your database. Make sure that the security group rules allow the appropriate incoming traffic.

○ **Choose existing**	● **Create new**
Choose existing VPC security groups	Create new VPC security group

New VPC security group name

bdok-database-sg

Availability Zone Info

No preference   ▼

Figure 10.13 – RDS subnet and security group configuration

9. Make sure that the **Database authentication** section is marked as **Password authentication**. All the other settings you can leave as default. At the end, AWS gives us the cost for this database if we keep it running for 30 days (*Figure 10.14*). Lastly, click **Create database** and wait a few minutes for the database creation:

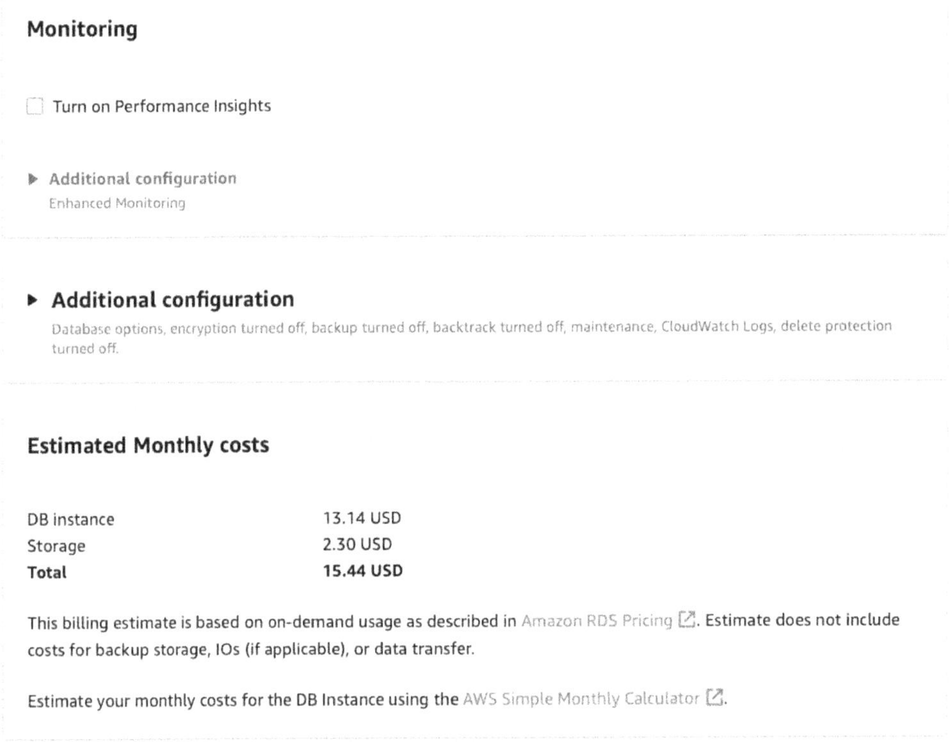

Monitoring

☐ Turn on Performance Insights

▶ Additional configuration
Enhanced Monitoring

▶ **Additional configuration**
Database options, encryption turned off, backup turned off, backtrack turned off, maintenance, CloudWatch Logs, delete protection
turned off.

Estimated Monthly costs

DB instance	13.14 USD
Storage	2.30 USD
Total	**15.44 USD**

This billing estimate is based on on-demand usage as described in Amazon RDS Pricing ☑. Estimate does not include
costs for backup storage, IOs (if applicable), or data transfer.

Estimate your monthly costs for the DB Instance using the AWS Simple Monthly Calculator ☑.

Figure 10.14 – Database estimate cost

10. Finally, we need to change the configurations for the database security group to allow connections
    from outside the VPC other than your own IP address (the default configuration). Go to the
    Databases page again and click on the newly created bdok-postgres database. Click on
    the security group name to open its page (*Figure 10.15*):

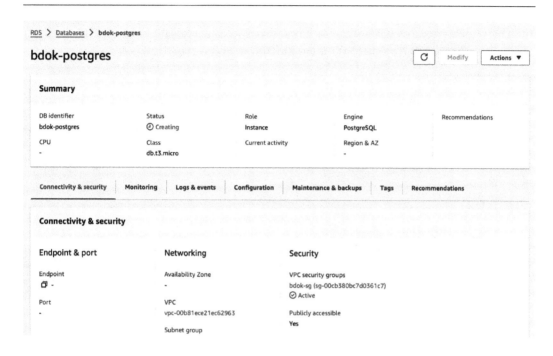

Figure 10.15 – bdok-postgres database view page

11. In the security group page, with the security group selected, roll down the page and click on the **Inbound rules** tab. Click on **Edit inbound rules** (*Figure 10.16*):

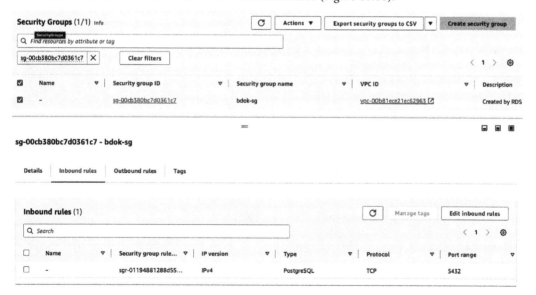

Figure 10.16 – Security group page

12. On the next page, you will see an entry rule already configured with your IP address as the source. Change it to **Anywhere-IPv4** and click on **Save rules** (*Figure 10.17*):

Figure 10.17 – Security rules configuration

13. Last thing – to populate our database with some data, we will use the `simulatinos.py` code to generate some fake customer data and ingest it into the database. The code is available at `https://github.com/PacktPublishing/Bigdata-on-Kubernetes/tree/main/Chapter10/streaming` folder. To run it, copy the database endpoint from its page on AWS, and in a terminal, type the following:

```
python simulations.py --host <YOUR-DATABASE-ENDPOINT> -p <YOUR-PASSWORD>
```

After the code prints some data on the terminal, stop the process with *Ctrl + C*. Now, we are set to start working on the data pipeline. Let's start with Kafka Connect configurations.

Deploying Kafka Connect and Elasticsearch

For Kafka to be able to access Elasticsearch, we will need to deploy another Elasticsearch cluster inside the same namespace Kafka is deployed. To do that, we will use two YAML configuration files, `elastic_cluster.yaml` and `kibana.yaml`. Both files are available in `https://github.com/PacktPublishing/Bigdata-on-Kubernetes/tree/main/Chapter10/streaming/elastic_deployment` folder. Follow the next steps:

1. First, download both files and run the following commands in a terminal:

```
kubectl apply -f elastic_cluster.yaml -n kafka
kubectl apply -f kibana.yaml -n kafka
```

2. Next, we will get an Elasticsearch automatically generated password with the following command:

```
kubectl get secret elastic-es-elastic-user -n kafka -o
go-template='{{.data.elastic | base64decode}}'
```

This command will print the password in the terminal. Save it for later.

3. Elasticsearch only works with encryption in transit. This means that we must configure certificates and keys that will allow Kafka Connect to correctly connect to Elastic. To that, first, we will get Elastic's certificates and keys and save them locally using the following commands:

```
kubectl get secret elastic-es-http-certs-public -n kafka
--output=go-template='{{index .data "ca.crt" | base64decode}}' >
ca.crt
kubectl get secret elastic-es-http-certs-public -n kafka
--output=go-template='{{index .data "tls.crt" | base64decode}}'
> tls.crt
kubectl get secret elastic-es-http-certs-internal -n kafka
--output=go-template='{{index .data "tls.key" | base64decode}}'
> tls.key
```

This will create three files locally, named `ca.crt`, `tls.crt`, and `tls.key`.

4. Now, we will use these files to create a `keystore.jks` file that will be used in the Kafka Connect cluster. In a terminal, run the following commands:

```
openssl pkcs12 -export -in tls.crt -inkey tls.key -CAfile ca.crt
-caname root -out keystore.p12 -password pass:BCoqZy82BhIhHv3C
-name es-keystore
keytool -importkeystore -srckeystore keystore.p12 -srcstoretype
PKCS12 -srcstorepass BCoqZy82BhIhHv3C -deststorepass
OfwxynZ8KATfZSZe -destkeypass OfwxynZ8KATfZSZe -destkeystore
keystore.jks -alias es-keystore
```

Note that I have set some random passwords. You can choose your own if you like. Now, you have the file we need to configure the encryption in transit, `keystore.jks`.

5. Next, we need to create a secret in Kubernetes using the `keystore.jks` file. To do this, in a terminal, type the following:

```
kubectl create secret generic es-keystore --from-file=keystore.
jks -n kafka
```

6. Now, we are ready to deploy Kafka Connect. We have a ready-to-go configuration file named `connect_cluster.yaml`, available at `https://github.com/PacktPublishing/Bigdata-on-Kubernetes/tree/main/Chapter10/streaming` folder. Two parts of this code, though, are worth mentioning. In *line 13*, we have the `spec.bootstrapServers` parameter. This parameter should be fulfilled with the service for Kafka bootstrap created by the Helm chart. To get the name of the service, type the following:

```
kubectl get svc -n kafka
```

Check if the service name matches the one in the code. If it doesn't, adjust accordingly. Keep the `9093` port for this service.

7. In *line 15*, you have the `spec.tls.trustedCertificates` parameter. The `secretName` value should match the exact name for the `ca-cert` secret created by the Helm chart. Check the name of this secret with the following command:

```
kubectl get secret -n kafka
```

If the name of the secret does not match, adjust accordingly. Keep the `ca.crt` value for the `certificate` parameter.

8. The last thing worth mentioning is that we will mount the `es-keystore` secret created before as a volume in Kafka Connect's pod. The following code block sets this configuration:

```
externalConfiguration:
  volumes:
    - name: es-keystore-volume
      secret:
        secretName: es-keystore
```

This secret must be mounted as a volume so that Kafka Connect can import the necessary secrets to connect to Elasticsearch.

9. To deploy Kafka Connect, in a terminal, type the following:

```
kubectl apply -f connect_cluster.yaml -n kafka
```

The Kafka Connect cluster will be ready in a couple of minutes. After it is ready, it is time to configure the **Java Database Connectivity** (**JDBC**) source connector to pull data from the Postgres database.

10. Next, prepare a YAML file that configures the JDBC source connector. Next, you will find the code for this file:

jdbc_source.yaml

```
apiVersion: "kafka.strimzi.io/v1beta2"
kind: "KafkaConnector"
metadata:
  name: "jdbc-source"
  namespace: kafka
  labels:
    strimzi.io/cluster: kafka-connect-cluster
spec:
  class: io.confluent.connect.jdbc.JdbcSourceConnector
  tasksMax: 1
  config:
    key.converter: org.apache.kafka.connect.json.JsonConverter
    value.converter: org.apache.kafka.connect.json.JsonConverter
    key.converter.schemas.enable: true
```

```
value.converter.schemas.enable: true
connection.url: «jdbc:postgresql://<DATABASE_ENDPOINT>:5432/
postgres»
connection.user: postgres
connection.password: "<YOUR_PASSWORD>"
connection.attempts: "2"
query: "SELECT * FROM public.customers"
mode: "timestamp"
timestamp.column.name: "dt_update"
topic.prefix: "src-customers"
valincrate.non.null: "false"
```

The connector's configuration specifies that it should use the `io.confluent.connect.jdbc.JdbcSourceConnector` class from Confluent's JDBC connector library. It sets the maximum number of tasks (parallel workers) for the connector to 1. The connector is configured to use JSON converters for both keys and values, with schema information included. It connects to a PostgreSQL database running on an Amazon RDS instance, using the provided connection URL, username, and password. The `SELECT * FROM public.customers` SQL query is specified, which means the connector will continuously monitor the `customers` table and stream out any new or updated rows as JSON objects in a Kafka topic named `src-customers`. The `mode` value is set to `timestamp`, which means the connector will use a timestamp column (`dt_update`) to track which rows have already been processed, avoiding duplicates. Finally, the `validate.non.null` option is set to `false`, which means the connector will not fail if it encounters `null` values in the database rows.

11. Place the YAML file in a folder named `connectors` and deploy the JDBC connector with the following command:

```
kubectl apply -f connectors/jdbc_source.yaml -n kafka
```

You can check if the connector was correctly deployed using the following command:

```
kubectl get kafkaconnector -n kafka
kubectl describe kafkaconnector jdbc-source -n kafka
```

We will also check if messages are correctly being delivered to the assigned Kafka topic using the following command:

```
kubectl exec kafka-cluster-kafka-0 -n kafka -c kafka -it -- bin/
kafka-console-consumer.sh --bootstrap-server localhost:9092
--from-beginning --topic src-customers
```

You should see the messages in JSON format printed on the screen. Great! We have a real-time connection with our source database. Now, it is time to set up the real-time processing layer with Spark.

Real-time processing with Spark

To correctly connect Spark with Kafka, we need to set up some authorization configuration in Kafka's namespace:

1. The following commands create a service account for Spark and set the necessary permissions to run `SparkApplication` instances in this environment:

   ```
   kubectl create serviceaccount spark -n kafka
   kubectl create clusterrolebinding spark-role-kafka
   --clusterrole=edit --serviceaccount=kafka:spark -n kafka
   ```

2. Next, we need to make sure that a secret with our AWS credentials is set in the namespace. Check if the secret already exists with the following command:

   ```
   kubectl get secrets -n kafka
   ```

   If the secret does not exist yet, create it with the following command:

   ```
   kubectl create secret generic aws-credentials --from-
   literal=aws_access_key_id=<YOUR_ACCESS_KEY_ID> --from-
   literal=aws_secret_access_key="<YOUR_SECRET_ACCESS_KEY>" -n
   kafka
   ```

3. Now, we need to build a Spark Streaming job. To do that, as seen before, we need a YAML configuration file and PySpark code that will be stored in Amazon S3. The YAML file follows the same pattern as seen before in *Chapter 8*. The code for this configuration is available at `https://github.com/PacktPublishing/Bigdata-on-Kubernetes/tree/main/Chapter10/streaming` folder. Save it locally as it will be used to deploy the `SparkApplication` job.

4. The Python code for the Spark job is also available in the GitHub repository under the `Chapter 10/streaming/processing` folder. It is named `spark_streaming_job.py`. This code is very similar to what we have seen in *Chapter 7*, but a few parts are worth commenting on. In *line 61*, we do real-time transformations on the data. Here, we are simply calculating the age of the person based on their birthdate using the following code:

   ```
   query = (
           newdf
           .withColumn("dt_birthdate", f.col("birthdate"))
           .withColumn("today", f.to_date(f.current_timestamp() ) )
           .withColumn("age", f.round(
               f.datediff(f.col("today"), f.col("dt_
   birthdate"))/365.25, 0)
               )
           .select("name", "gender", "birthdate", "profession",
   "age", "dt_update")
       )
   ```

5. For the Elasticsearch sink connector to read messages on topics correctly, the messages must be in a standard Kafka JSON format with two keys: `schema` and `payload`. In the code, we will manually build this schema and concatenate it to the final version of the data in JSON format. *Line 70* defines the `schema` key and the beginning of the `payload` structure (the line will not be printed here to improve readability).

6. In *line 72*, we transform the values of the DataFrame into a single JSON string and set it to a column named `value`:

```
json_query = (
    query
    .select(
        f.to_json(f.struct(f.col("*")))
    )
    .toDF("value")
)
```

7. In *line 80*, we concatenate the previously defined `schema` key for the JSON string with the actual values of the data and write it in a streaming query back to Kafka in a topic named `customers-transformed`:

```
(
    json_query
    .withColumn("value", f.concat(f.lit(write_schema),
f.col("value"), f.lit('}')))
    .selectExpr("CAST(value AS STRING)")
    .writeStream
    .format("kafka")
    .option("kafka.bootstrap.servers", "kafka-cluster-kafka-
bootstrap:9092")
    .option("topic", "customers-transformed")
    .option("checkpointLocation", "s3a://bdok-<ACCOUNT-
NUMBER>/spark-checkpoint/customers-processing/")
    .start()
    .awaitTermination()
)
```

8. Save this file as `spark_streaming_job.py` and save it in the S3 bucket we defined in the YAML file. Now, you are ready to start the real-time processing. To start the streaming query, in a terminal, type the following command:

```
kubectl apply -f spark_streaming_job.yaml -n kafka
```

You can also check if the application is running correctly with the following commands:

```
kubectl describe sparkapplication spark-streaming-job -n kafka
kubectl get pods -n kafka
```

9. Now, check if the messages are being correctly written into the new topic with the following:

```
kubectl exec kafka-cluster-kafka-0 -n kafka -c kafka -it -- bin/
kafka-console-consumer.sh --bootstrap-server localhost:9092
--from-beginning --topic customers-transformed
```

That's it! We have the real-time processing layer up and running. Now, it is time to deploy the Elasticsearch sink connector and get the final data into Elastic. Let's get to it.

Deploying the Elasticsearch sink connector

Here, we will begin with the YAML configuration file for the Elasticsearch sink connector. Most of the "heavy lifting" was done earlier with the configuration of the secrets needed:

1. Create a file named es_sink.yaml under the connectors folder. Here is the code:

es_sink.yaml

```
apiVersion: "kafka.strimzi.io/v1beta2"
kind: "KafkaConnector"
metadata:
  name: "es-sink"
  namespace: kafka
  labels:
    strimzi.io/cluster: kafka-connect-cluster
spec:
  class: io.confluent.connect.elasticsearch.
ElasticsearchSinkConnector
  tasksMax: 1
  config:
    topics: "customers-transformed"
    connection.url: "https://elastic-es-http.kafka:9200"
    connection.username: "elastic"
    connection.password: "w6MR9V0SNLD79b56arB9Q6b6"
    batch.size: 1
    key.ignore: "true"
    elastic.security.protocol: "SSL"
    elastic.https.ssl.keystore.location: "/opt/kafka/external-
configuration/es-keystore-volume/keystore.jks"
    elastic.https.ssl.keystore.password: "OfwxynZ8KATfZSZe"
    elastic.https.ssl.key.password: "OfwxynZ8KATfZSZe"
    elastic.https.ssl.keystore.type: "JKS"
    elastic.https.ssl.truststore.location: "/opt/kafka/external-
configuration/es-keystore-volume/keystore.jks"
    elastic.https.ssl.truststore.password: "OfwxynZ8KATfZSZe"
    elastic.https.ssl.truststore.type: "JKS"
```

The part I think is worth some attention here is from *line 20* on. Here, we are configuring the SSL/TLS settings for the connection to Elasticsearch. The `keystore.location` and `truststore.location` properties specify the paths to the `keystore` and `truststore` files, respectively (which, in this case, are the same). The `keystore.password`, `key.password`, and `truststore.password` properties provide the passwords for accessing these files. The `keystore.type` and `truststore.type` properties specify the type of the `keystore` and `truststore` files, which in this case is JKS (Java KeyStore).

2. Now, everything is set to get this connector up and running. In a terminal, type the following:

```
kubectl apply -f connectors/es_sink.yaml -n kafka
```

You can also check if the connector was correctly deployed with the following command:

```
kubectl describe kafkaconnector es-sink -n kafka
```

3. Now, get the load balancer's URL and access the Elasticsearch UI. Let's see if our data got correctly ingested:

```
kubectl get svc -n kafka
```

4. Once you are logged in to Elasticsearch, choose **Dev Tools** in the side menu and run the `GET _cat/indices` command. If all is well, the new `customers-transformed` index will show up in the output (*Figure 10.18*):

Figure 10.18 – New index created in Elasticsearch

5. Now, let's create a new data view with this index. In the side menu, choose **Stack Management** and click on **Data Views**. Click on the **Create data view** button.

6. Set `customers-transformed` for the data view name and again `customers-transformed` for the index pattern. Select the `dt_update` column as the timestamp field. Then, click on **Save data view to Kibana** (*Figure 10.19*):

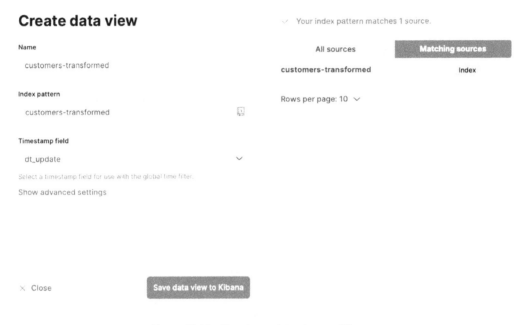

Figure 10.19 – Creating a data view on Kibana

7. Now, let's check the data. In the side menu, choose **Discover** and then select the newly created `customers-transformed` data view. Remember to set the date filter to a reasonable value, such as 1 year ago. You can play with some time-based subsets of the data if you are doing a larger indexing. The data should be shown in the UI (*Figure 10.20*):

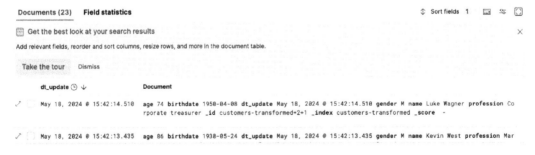

Figure 10.20 – customers-transformed data shown in Kibana

Now, add more data by running the `simulations.py` code again. Try to play a little bit and build some cool dashboards to visualize your data.

And that is it! You just ran an entire real-time data pipeline in Kubernetes using Kafka, Spark, and Elasticsearch. Cheers, my friend!

Summary

In this chapter, we brought together all the knowledge and skills acquired throughout the book to build two complete data pipelines on Kubernetes: a batch processing pipeline and a real-time pipeline. We started by ensuring that all the necessary tools, such as a Spark operator, a Strimzi operator, Airflow, and Trino, were correctly deployed and running in our Kubernetes cluster.

For the batch pipeline, we orchestrated the entire process, from data acquisition and ingestion into a data lake on Amazon S3 to data processing using Spark, and finally delivering consumption-ready tables in Trino. We learned how to create Airflow DAGs, configure Spark applications, and integrate different tools seamlessly to build a complex, end-to-end data pipeline.

In the real-time pipeline, we tackled the challenges of processing and analyzing data streams in real time. We set up a Postgres database as our data source, deployed Kafka Connect and Elasticsearch, and built a Spark Streaming job to perform real-time transformations on the data. We then ingested the transformed data into Elasticsearch using a sink connector, enabling us to build applications that can react to events as they occur.

Throughout the chapter, we gained hands-on experience in writing code for data processing, orchestration, and querying using Python and SQL. We also learned best practices for integrating different tools, managing Kafka topics, and efficiently indexing data into Elasticsearch.

By completing the exercises in this chapter, you have acquired the skills to deploy and orchestrate all the necessary tools for building big data pipelines on Kubernetes, connect these tools to successfully run batch and real-time data processing pipelines, and understand and apply best practices for building scalable, efficient, and maintainable data pipelines.

In the next chapter, we will discuss how we can use Kubernetes to deploy **generative AI (GenAI)** applications.

11

Generative AI on Kubernetes

Generative artificial intelligence (GenAI) has emerged as a transformative technology, revolutionizing how we interact with and leverage AI. In this chapter, we will explore the exciting world of generative AI and learn how to harness its power on Kubernetes. We will dive into the fundamentals of generative AI and understand its main differences from traditional AI.

Our focus will be on leveraging **Amazon Bedrock**, a comprehensive suite of services designed to simplify the development and deployment of generative AI applications. Through hands-on examples, you will gain practical experience in building a generative AI application on Kubernetes using **Streamlit**, a powerful Python library for creating interactive data applications. We will cover the entire process, from the development to deploying the application on a Kubernetes cluster.

Moreover, we will explore the concept of **retrieval-augmented generation (RAG)**, which combines the power of generative AI with external knowledge bases.

Finally, we will introduce **Agents for Amazon Bedrock**, a powerful feature that allows you to automate tasks and create intelligent assistants. You will learn how to build an agent, define its capabilities through an OpenAPI schema, and create the underlying Lambda function that serves as the backend for your agent.

By the end of this chapter, you will have a solid understanding of generative AI, its applications, and the tools and techniques required to build and deploy generative AI applications on Kubernetes.

In this chapter, we're going to cover the following main topics:

- What generative AI is and what it is not
- Using Amazon Bedrock to work with foundational models
- Building a generative AI application on Kubernetes
- Building RAG with **Knowledge Bases for Amazon Bedrock**
- Building action models with agents

Technical requirements

For this chapter, you will need an AWS account and a running Kubernetes cluster. We will also be using **LangChain** and Streamlit libraries. Although it is not necessary to have them installed for application deployment in Kubernetes, the installation is advised if you want to test the code locally and modify it to your own experiments.

Also, it will be necessary to install the **Beautiful Soup** library to get data for the RAG exercise (fourth section).

All the code for this chapter is available at `https://github.com/PacktPublishing/Bigdata-on-Kubernetes` under the `Chapter11` folder.

What generative AI is and what it is not

At its core, generative AI refers to AI systems capable of generating new, original content, such as text, images, audio, or code, based on the training data they have been exposed to. Generative AI models are trained on large datasets of existing content, and they learn the patterns and relationships within that data. When prompted, these models can then generate new, original content that resembles the training data but is not an exact copy of any specific example.

This contrasts with traditional machine learning models, which are focused on making predictions or classifications based on existing data.

Traditional machine learning models, such as those used for image recognition, natural language processing, or predictive analytics, are designed to take in input data and make predictions or classifications based on that data. Machine learning models excel at tasks such as classification (e.g., identifying objects in images or topics in texts), regression (e.g., predicting house prices based on features such as square footage and location), and clustering (e.g., grouping customers based on similar behavior patterns).

For example, an image recognition model might be trained on a large dataset of labeled images to learn to recognize and classify objects in new, unseen images. Similarly, a natural language processing model might be trained on a corpus of text data to perform tasks such as sentiment analysis, named entity recognition, or language translation.

In a credit risk assessment scenario, a machine learning model would be trained on a dataset containing information about past loan applicants, such as their income, credit history, and other relevant features, along with labels indicating whether they defaulted on their loans or not. The model would learn the patterns and relationships between these features and the loan default outcomes. When presented with a new loan application, the trained model can then predict the likelihood of the applicant defaulting on the loan.

In these cases, the machine learning model is not generating new content; instead, it is using the patterns and relationships it has learned from the training data to make informed predictions or decisions about new, unseen data.

In contrast, for instance, a generative AI model trained on a vast corpus of text can generate human-like writing on any given topic or in any desired style. Similarly, models trained on images can create entirely new, realistic-looking images based on textual descriptions or other input data.

While the end result of generative AI is the creation of new content, the underlying mechanism is still based on the same principles of machine learning: making predictions. However, instead of predicting a single output (such as a classification or a numerical value), generative AI models are trained to predict the next element in a sequence, whether that sequence is a sequence of words, pixels, or any other type of data.

The power of large neural networks

While the concept of predicting the next element in a sequence is relatively simple, the ability of generative AI models to generate coherent, high-quality content lies in the sheer scale and complexity of the neural networks used to power these models.

Generative AI models typically employ large, deep neural networks with billions or even trillions of parameters. These neural networks are trained on vast amounts of data, often spanning millions or billions of examples, allowing them to capture incredibly nuanced patterns and relationships within the data.

For example, the Anthropic models, such as Claude, are trained on an enormous corpus of text data, spanning a wide range of topics and domains. This allows the models to develop a deep understanding of language, context, and domain-specific knowledge, enabling them to generate text that is not only grammatically correct but also semantically coherent and relevant to the given context.

Challenges and limitations

While generative AI has demonstrated remarkable capabilities, it is not without its challenges and limitations. One of the primary concerns is the potential for these models to generate biased, harmful, or misleading content, especially when trained on datasets that reflect societal biases or contain inaccurate information.

Additionally, generative AI models can sometimes produce outputs that are nonsensical, inconsistent, or factually incorrect, even though they may appear coherent and plausible on the surface. This is known as the "hallucination" problem, where the model generates content that is not grounded in factual knowledge or the provided context. Here are two well-known real-life cases. Air Canada's AI-powered chatbot provided misleading information to a passenger regarding the airline's bereavement fare policy. The chatbot incorrectly stated that passengers could apply for reduced bereavement fares retroactively, even after travel had already occurred, which contradicted Air Canada's actual policy. The passenger relied on this hallucinated response from the chatbot and subsequently filed a successful small claims case against Air Canada when the airline refused to honor the chatbot's advice (`https://www.forbes.com/sites/marisagarcia/2024/02/19/what-air-canada-lost-in-remarkable-lying-ai-chatbot-case/`). Also, a federal judge in Brazil used the ChatGPT AI system to research legal precedents for a ruling he was writing. However, the AI provided fabricated information, citing non-existent rulings from the Superior Court of Justice as the basis for the judge's decision (`https://g1.globo.com/politica/blog/daniela-lima/post/2023/11/13/juiz-usa-inteligencia-artificial-para-fazer-decisao-e-cita-jurisprudencia-falsa-cnj-investiga-caso.ghtml`).

Despite these challenges, generative AI is a rapidly evolving field, and researchers and developers are actively working on addressing these issues. Techniques such as fine-tuning, prompt engineering, and the use of external knowledge sources (e.g., knowledge bases or RAG) are being explored to improve the reliability, safety, and factual accuracy of generative AI models.

In the following sections, we will dive deeper into the practical aspects of building and deploying generative AI applications using Amazon Bedrock and its foundational models, knowledge base, and agent-based architectures.

Using Amazon Bedrock to work with foundational models

Amazon Bedrock provides a suite of foundational models that can be used as building blocks for your generative AI applications. It's important to understand the capabilities and intended use cases of each model to choose the right one for your application.

The available models in Amazon Bedrock include language models, computer vision models, and multimodal models. Language models excel at understanding and generating human-like text. They can be employed for tasks such as text summarization, question answering, and content generation. Computer vision models, on the other hand, are adept at analyzing and understanding visual data, making them ideal for applications such as image recognition, object detection, and scene understanding.

Multimodal models, as the name suggests, can handle multiple modalities simultaneously. This makes it suitable for tasks such as image captioning, visual question answering, and data chart analysis.

It's important to note that each model has its own strengths and limitations, and the choice of model should be guided by the specific requirements of your application. For example, if your application primarily deals with text-based tasks, a language model such as Llama might be the most appropriate choice. However, if you need to process both text and images, a multimodal model such as Claude would be a better fit.

To effectively integrate Amazon Bedrock's foundational models into our generative AI applications, follow these steps:

1. To use Amazon Bedrock's available foundational models, first, we need to activate them. Go to the AWS console and search for the Amazon Bedrock page. Then, click on **Modify model access** (*Figure 11.1*).

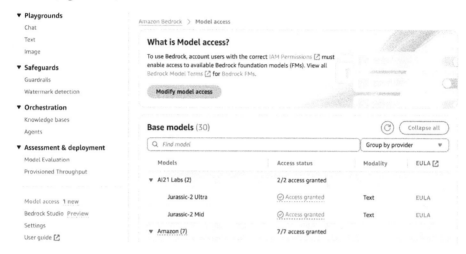

Figure 11.1 – Modifying model access on Amazon Bedrock

2. On the next page, select the **Claude 3 Sonnet** and **Claude 3 Haiku** Anthropic models. Those are the foundational models we will use for our generative AI applications. You can select all the available models if you wish to play and experiment with different models (*Figure 11.2*).

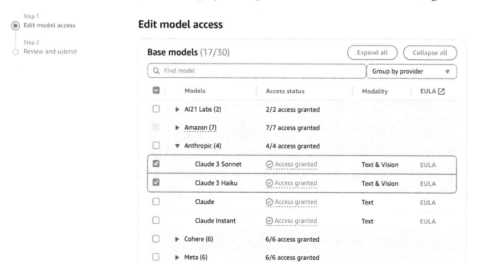

Figure 11.2 – Requesting access for Anthropic's Claude 3 models

3. Click **Next** and, on the next page, review the changes and click **Submit**. Those models can take a few minutes to get access granted.

Once access has been granted, we have all we need to develop a generative AI application. Let's get to it.

Building a generative AI application on Kubernetes

In this section, we will build a generative AI application with Streamlit. A diagram representing the architecture for this application is shown in *Figure 11.3*. In this application, the user will be able to choose which foundational model they are going to talk to.

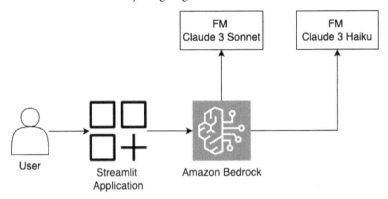

Figure 11.3 – Foundational models' application architecture

Let's start with the Python code for the application. The complete code is available under the `Chapter 11/streamlit-claude/app` folders on GitHub. We will walk through the code, block by block:

1. Create a folder named `app` and inside it, create a `main.py` code file. First, we import the necessary files and create a client to access Amazon Bedrock runtime APIs:

```
import boto3
from langchain_community.chat_models import BedrockChat
from langchain.callbacks.streaming_stdout import
StreamingStdOutCallbackHandler

bedrock = boto3.client(service_name='bedrock-runtime', region_
name="us-east-1")
```

2. Next, we define a dictionary of parameters that are important for working with Claude:

```
inference_modifier = {
    "max_tokens": 4096,
    "temperature": 0.5,
    "top_k": 250,
```

```
    "top_p": 1,
    "stop_sequences": ["\n\nHuman:"],
}
```

3. Next, we will configure a function to allow the choosing of the preferred foundational model. With the choice, we will return a model object that can access Bedrock through Langchain:

```python
def choose_model(option):
    modelId = ""
    if option == "Claude 3 Haiku":
        modelId = "anthropic.claude-3-haiku-20240307-v1:0"
    elif option == "Claude 3 Sonnet":
        modelId = "anthropic.claude-3-sonnet-20240229-v1:0"

    model = BedrockChat(
        model_id=modelId,
        client=bedrock,
        model_kwargs=inference_modifier,
        streaming=True,
        callbacks=[StreamingStdOutCallbackHandler()],
    )
    return model
```

4. Now, we will add a small function to reset the conversation history:

```python
def reset_conversation():
    st.session_state.messages = []
```

5. Next, we will begin the development of the `main` function and add some widgets to the application interface. The following code creates a sidebar. In it, we add a selection box with Claude 3 Haiku and Claude 3 Sonnet as options, we write a confirmation message to tell the user which model they are talking to, and we add a **Reset Chat** button. After that, we run the `choose_model` function to return the class that connects to Bedrock and write the title of the application, *Chat with Claude 3*:

```python
def main():
    with st.sidebar:
        option = st.selectbox(
            "What model do you want to talk to?",
            ("Claude 3 Haiku", "Claude 3 Sonnet")
        )
        st.write(f"You are talking to **{option}**")

        st.button('Reset Chat', on_click=reset_conversation)
```

```
model = choose_model(option)

st.title("Chat with Claude 3")
```

6. Next, we will initialize the chat history as an empty list if it doesn't already exist in `st.session_state`. `st.session_state` is a Streamlit object that persists data across app reruns. Then, we iterate over the `messages` list in `st.session_state` and display each message in a chat message container. The `st.chat_message` function creates a chat message container with the specified role (e.g., `user` or `assistant`). The `st.markdown` function displays the message content inside the container:

```
if "messages" not in st.session_state:
    st.session_state.messages = []

for message in st.session_state.messages:
    with st.chat_message(message["role"]):
        st.markdown(message["content"])
```

7. Next, we handle user input and display the conversation. The `st.chat_input` function creates an input field where the user can enter their prompt. If the user enters a prompt, the following steps are executed: (1) the user's prompt is added to the `messages` list in `st.session_state` with the `user` role; (2) the user's prompt is displayed in a chat message container with the `user` role; (3) the `model.stream(prompt)` function is called, which sends the user's prompt to the Bedrock model and streams the response back. The `st.write_stream` function displays the streamed response in real-time; (4) the assistant's response is added to the `messages` list in `st.session_state` with the `assistant` role:

```
if prompt := st.chat_input("Enter your prompt here"):
    st.session_state.messages.append(
        {"role": "user", "content": prompt}
    )
    with st.chat_message("user"):
        st.markdown(prompt)

    with st.chat_message("assistant"):
        response = st.write_stream(
            model.stream(prompt)
        )
    st.session_state.messages.append(
        {"role": "assistant", "content": response}
    )
```

8. Finally, we call the main function to start the Streamlit application:

```
if __name__ == "__main__":
    main()
```

If you want to run this application locally, here is a `requirements.txt` file:

```
boto3==1.34.22
langchain-community==0.0.33
langchain==0.1.16
streamlit==1.34.0
```

Install the necessary libraries with the following:

```
pip install -r requirements.txt
```

If you have the libraries already installed, authenticate your AWS CLI with the `aws configure` command and start the application locally with the following:

```
streamlit run main.py
```

This is an awesome way of testing the application before building a container image for deployment. You can test and modify the application as you wish.

When it is ready, now, let's build a container image for deployment.

9. The following is a simple **Dockerfile** to build the image:

Dockerfile

```
FROM python:3.9-slim

WORKDIR /app

RUN apt-get update && apt-get install -y \
    build-essential \
    curl \
    software-properties-common \
    git \
    && rm -rf /var/lib/apt/lists/*

COPY app /app/

EXPOSE 8501

HEALTHCHECK CMD curl --fail http://localhost:8501/_stcore/health

RUN pip3 install -r requirements.txt

ENTRYPOINT ["streamlit", "run", "main.py", "--server.port=8501",
"--server.address=0.0.0.0"]
```

This Dockerfile starts with the Python 3.9 slim base image and sets the working directory to /app. It then installs various system packages required for the application, such as `build-essential`, `curl`, `software-properties-common`, and `git`. The application code is copied into the /app directory, and the container exposes port 8501. A health check is set up to check whether the Streamlit application is running correctly on http://localhost:8501/_stcore/health. The required Python packages are installed using pip3 based on the `requirements.txt` file. Finally, the ENTRYPOINT command starts the Streamlit application by running `streamlit run main.py` and specifying the server port and address.

10. To build the image locally, type the following:

```
docker build --platform linux/amd64 -t <YOUR_USERNAME>/chat-with-claude:v1 .
```

Remember to change <YOUR_USERNAME> to your actual Docker Hub username. Then, push the image with the following:

```
docker push <YOUR_USERNAME>/chat-with-claude:v1
```

Remember that this image is going to be publicly available on Docker Hub. Don't put any authentication credentials or sensitive data in the code or as environment variables!

Now, let's deploy our application on Kubernetes.

Deploying the Streamlit app

As we have seen before, to deploy our app on Kubernetes, we need a Deployment and a Service .yaml definition. We can provide both in a single file:

1. First, create a `deploy_chat_with_claude.yaml` file with this code:

deploy_chat_with_claude.yaml

```
apiVersion: apps/v1
kind: Deployment
metadata:
  name: chat-with-claude
spec:
  replicas: 1
  selector:
    matchLabels:
      app: chat-with-claude
  template:
    metadata:
      labels:
        app: chat-with-claude
```

```
spec:
  containers:
  - name: chat-with-claude
    image: docker.io/neylsoncrepalde/chat-with-claude:v1
    ports:
    - containerPort: 8501
    env:
    - name: AWS_ACCESS_KEY_ID
      valueFrom:
        secretKeyRef:
          name: aws-credentials
          key: aws_access_key_id
    - name: AWS_SECRET_ACCESS_KEY
      valueFrom:
        secretKeyRef:
          name: aws-credentials
          key: aws_secret_access_key
```

The first part of the code defines a `Deployment` resource named `chat-with-claude`. It takes a previously built image (which you can change to your own new image) and opens port `8501` in the container to be accessed from outside the pod. The `spec.template.spec.containers.env` block mounts AWS credentials as environment variables in the container from a secret called `aws-credentials`.

2. The second part of the code defines a `LoadBalancer` service for the pods defined in `Deployment`, which listens on port `8501` and directs traffic to port `8501` in the container. Don't forget `---`, which is necessary to separate several resources in a single file:

```
---
apiVersion: v1
kind: Service
metadata:
  name: chat-with-claude
spec:
  type: LoadBalancer
  ports:
  - port: 8501
    targetPort: 8501
  selector:
    app: chat-with-claude
```

3. Now, we are going to create the namespace and the secret and deploy the application with the following:

```
kubectl create namespace genai
kubectl create secret generic aws-credentials --from-
literal=aws_access_key_id=<YOUR_ACCESS_KEY_ID> --from-
literal=aws_secret_access_key="<YOUR_SECRET_ACCESS_KEY>" -n
genai
kubectl apply -f deploy_chat_with_claude.yaml -n genai
```

4. That's it. Wait a few minutes for `LoadBalancer` to be up and running and check its URL with the following:

```
kubectl get svc -n genai
```

5. Now, paste the URL with `:8501` at the end to define the correct port *et voilà!* (*Figure 11.4*).

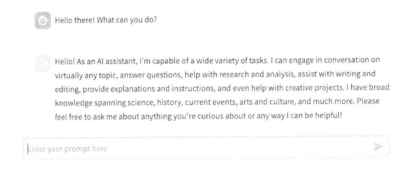

Figure 11.4 – The Chat with Claude 3 app UI

Now, play a little bit with the assistant. Try Haiku and Sonnet and note their differences in speed and quality of answer. After a few shots, you will notice that asking specific questions to foundational models leads to a hallucination. Ask the model, for instance, who are you. You are going to have a nice surprise (and some laughs). This model needs context.

In the next section, we will provide some context using RAG.

Building RAG with Knowledge Bases for Amazon Bedrock

RAG is a technique used in generative AI models to provide additional context and knowledge to foundational models during the generation process. It works by first retrieving relevant information from a knowledge base or corpus of documents, and then using this retrieved information to augment the input to the generative model.

RAG is a good choice for giving context to generative AI models because it allows the model to access and utilize external knowledge sources, which can significantly improve the quality, accuracy, and relevance of the generated output. Without RAG, the model would be limited to the knowledge and patterns it learned during training, which may not always be sufficient or up to date, especially for domain-specific or rapidly evolving topics.

One of the key advantages of RAG is that it enables the model to leverage large knowledge bases or document collections, which would be impractical or impossible to include in the model's training data. This allows the model to generate more informed and knowledgeable outputs, as it can draw upon a vast amount of relevant information. Additionally, RAG can help mitigate issues such as hallucination and bias, as the model has access to authoritative and factual sources.

However, RAG also has some limitations. The quality of the generated output heavily depends on the relevance and accuracy of the retrieved information, which can be influenced by the quality of the knowledge base, the effectiveness of the retrieval mechanism, and the ability of the model to properly integrate the retrieved information. Additionally, RAG can introduce computational overhead and latency, as it requires an additional retrieval step before the generation process.

To build an AI assistant with RAG, we will use the Knowledge Bases for Amazon Bedrock service, a feature in Bedrock that allows you to create and manage a knowledge base seamlessly. Let's get to it.

For our exercise, we will build an AI assistant capable of giving information about the AWS Competency Program. A visual representation of this assistant's architecture is shown in *Figure 11.5*:

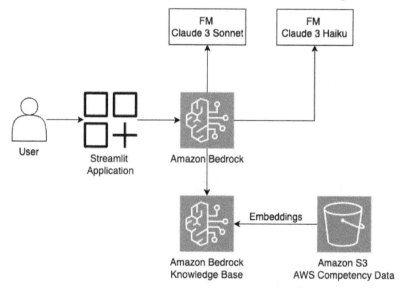

Figure 11.5 – Knowledge Bases for Amazon Bedrock application architecture

The AWS Competency Program is a validation program offered by AWS that recognizes partners who have demonstrated technical proficiency and proven customer success in specialized solution areas. AWS Competencies are awarded to **AWS Partner Network** (**APN**) members who have undergone technical validation related to specific AWS services or workloads, ensuring they have the expertise needed to deliver consistent, high-quality solutions on AWS. These competencies span various areas such as DevOps, migration, data and analytics, machine learning, and security. Each competency has its own rules document and can be quite challenging to understand.

1. First, we will gather some context information about the program. On GitHub, under the `Chapter 11/claude-kb/knowledge-base/` folder, you will find a Python code that will gather information on conversational AI, data and analytics, DevOps, education, energy, financial services, machine learning, and security programs. After saving this code locally, install the Beautiful Soup library with the following:

   ```
   pip install "beautifulsoup4==4.12.2"
   ```

   Then, run the code with the following:

   ```
   python get_competency_data.py
   ```

   After a few seconds, data should be saved locally on your machine.

2. Next, create an S3 bucket and upload these files. This will be the base for our RAG layer.

3. Next, go to the **Bedrock** page in the AWS console. In the side menu, click on **Knowledge Bases** and then, click on **Create knowledge base** (*Figure 11.6*).

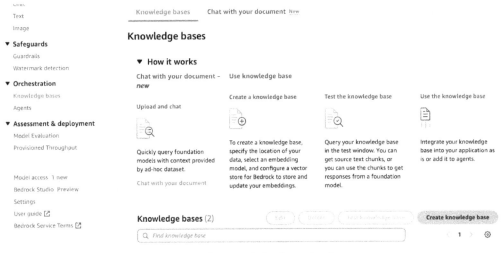

Figure 11.6 – The Knowledge bases landing page

4. On the next page, choose a name for your knowledge base and select **Create and use a new service role** under the **IAM permissions** section. Then, click **Next**.

5. Next, you will configure the data source. Choose a data source name as you wish. For **Data source location**, make sure the **This AWS account** box option is checked. Then, in the **S3 URI** section, click on **Browse S3** to search for your S3 bucket that contains the AWS Competency datasets (the bucket we created in *Step 2*). An example of that configuration is shown in *Figure 11.7*. After selecting the S3 bucket, click **Next**.

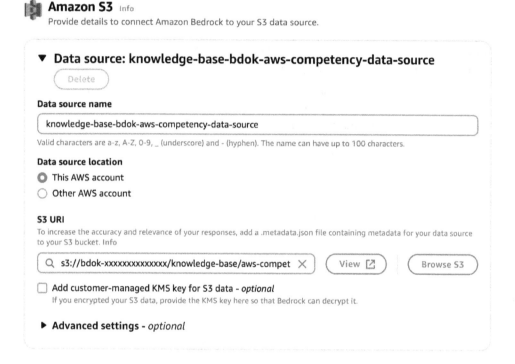

Figure 11.7 – Choosing a data source for the knowledge base

Next, we are going to choose the embeddings model. This embeddings model is responsible for transforming text or image files into vector representations called **embeddings**. These embeddings capture the semantic and contextual information of the input data, allowing for efficient similarity comparisons and retrieval operations. One of Bedrock's embeddings models, Amazon Titan, should be available by default. If it is not, do the same process of asking for access in the console.

6. On the next page, in the **Embeddings model** section, choose **Titan Embeddings G1 - Text**. In the **Vector database** section, make sure the **Quick create a new vector store** option is checked. This quick creation option creates a vector database based on OpenSearch Serverless. Leave the other options unmarked and click **Next**.

> **Note**
>
> OpenSearch is an open-source distributed search and analytics engine based on Apache Lucene and derived from Elasticsearch. It is a great option for a RAG vector database because it provides efficient full-text search and nearest-neighbor search capabilities for vector embeddings. OpenSearch supports dense vector indexing and retrieval, making it suitable for storing and querying large collections of vector embeddings, which are essential for the retrieval component of RAG models.

7. Next, review the information to see whether it was correctly provided. If everything looks good, click on **Create knowledge base**. Be patient. This creation will take several minutes to complete.

8. After the knowledge base is up and running, go back to the **Knowledge base** page in Bedrock and click on the knowledge base you just created. On the next page, scroll until you find the **Data source** section (as shown in *Figure 11.8*). Select the data source and click **Sync** to start the embedding of the text content. This will also take a few minutes.

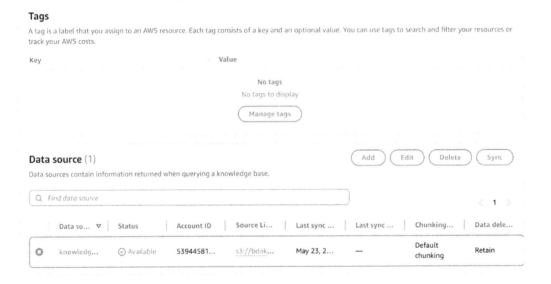

Figure 11.8 – Syncing the knowledge base with its data source

After the "sync" is ready, we have everything we need to run our generative AI assistant with RAG. Now, it is time to adjust the code to let Claude work with the knowledge base.

Adjusting the code for RAG retrieval

We will start from the code we developed earlier to work with the pure Claude model. As we just need some small modifications, we won't go through the entire code again. We will take a closer look at the necessary modifications. The complete code for the RAG application is available at `https://github.com/PacktPublishing/Bigdata-on-Kubernetes/tree/main/Chapter11/claude-kb/app` folder. If you don't want to customize your code, you can use the ready-to-go docker image I provided for this example.

1. First, we need extra imports:

    ```
    import os
    from botocore.client import Config
    from langchain.prompts import PromptTemplate
    from langchain.retrievers.bedrock import
    AmazonKnowledgeBasesRetriever
    from langchain.chains import RetrievalQA
    ```

    Here, we import the `os` library to get environment variables. The `Config` class will help build a configuration object to access the `bedrock-agent` API. All the other imports relate to accessing the knowledge base and merging the retrieved documents with AI responses.

2. Next, we will get the ID for the Knowledge Bases for Amazon Bedrock service from an environment variable. This can be a very helpful approach. If we need to change the knowledge base in the future, there is no need to rebuild the image. We just change the environment variable. Then, we set some configurations and create a client for the `bedrock-agent-runtime` API (needed for the knowledge base):

    ```
    kb_id = os.getenv("KB_ID")

    bedrock_config = Config(connect_timeout=120, read_timeout=120,
    retries={'max_attempts': 0})

    bedrock_agent_client = boto3.client(
        "bedrock-agent-runtime", config=bedrock_config, region_name
    = "us-east-1"
    )
    ```

3. Next, we will configure a prompt template that will help us chain the retrieved documents from the knowledge base and the user questions. At the end, we instantiate an object that will hold the template and receive the documents and the user questions as inputs:

    ```
    PROMPT_TEMPLATE = """
    Human: You are a friendly AI assistant and provide answers to
    questions about AWS competency program for partners.
    Use the following pieces of information to provide a concise
    answer to the question enclosed in <question> tags.
    Don't use tags when you generate an answer. Answer in plain
    text, use bullets or lists if needed.
    If you don't know the answer, just say that you don't know,
    don't try to make up an answer.
    ```

```
<context>
{context}
</context>

<question>
{question}
</question>

The response should be specific and use statistics or numbers
when possible.

Assistant:"""

claude_prompt = PromptTemplate(template=PROMPT_TEMPLATE,
                                    input_
variables=["context","question"])
```

4. After setting the `choose_model()` function, we need to instantiate a `retriever` class that will pull documents from the knowledge base:

```
retriever = AmazonKnowledgeBasesRetriever(
        knowledge_base_id=kb_id,
        retrieval_config={
            "vectorSearchConfiguration": {
                "numberOfResults": 4
            }
        },
        client=bedrock_agent_client
    )
```

5. Now, inside the `main` function, we will add `RetrievalQA`. This class is used for building question-answering systems that can retrieve relevant information from the knowledge base:

```
qa = RetrievalQA.from_chain_type(
        llm=model,
        chain_type="stuff",
        retriever=retriever,
        return_source_documents=False,
        chain_type_kwargs={"prompt": claude_prompt}
    )
```

6. Finally, we will modify the response to give the entire answer:

```
with st.chat_message("assistant"):
    response = qa.invoke(prompt)['result']
    st.write(response)
```

That's it. The code is ready to be built in a new image. You can rebuild it by creating a new Dockerfile with the same code we used before. When running the `docker build` command, remember to choose a different image name (or, at least, a different version).

7. Next, we will start the deployment. The `.yaml` file is also very similar to the one we did in the last section (but remember to change all the names for the deployment, services, container, and label to `rag-with-claude`). A full version of this code is available in the GitHub repository). We only need to declare the environment variable for the knowledge base ID. As this is not a sensitive credential, we don't need to use a Kubernetes secret for that. We will use `ConfigMap`. The `spec.template.spec.container.env` section of your `.yaml` file should look like this:

```
env:
  - name: AWS_ACCESS_KEY_ID
    valueFrom:
      secretKeyRef:
        name: aws-credentials
        key: aws_access_key_id
  - name: AWS_SECRET_ACCESS_KEY
    valueFrom:
      secretKeyRef:
        name: aws-credentials
        key: aws_secret_access_key
  - name: KB_ID
    valueFrom:
      configMapKeyRef:
        name: kb-config
        key: kb_id
```

Note that we added a new environment variable called `KB_ID` that will be imported from `ConfigMap`.

8. To deploy the new application, we run the following:

```
kubectl create configmap kb-config --from-literal=kb_id=<YOUR_
KB_ID> -n genai
```

We run the preceding to create `ConfigMap` with the knowledge base ID, and then we run:

```
kubectl apply -f deploy_chat_with_claude.yaml -n genai
```

We run the preceding to deploy the application. Wait a few minutes for `LoadBalancer` to be up and use the following command:

```
kubectl get svc -n genai
```

Use the preceding command to get the URL for `LoadBalancer`. Copy and paste the service named `rag-with-claude` in a browser and add `:8501` to connect to the exposed port. *Et voilà!* You should see your new application running as shown in *Figure 11.9*.

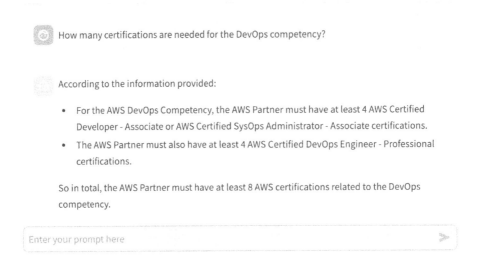

Figure 11.9 – RAG application UI

Try to play a little bit with this application. You will see that if you ask questions not related to its scope (AWS Competency program), the assistant will say it cannot answer.

Now, we will move to the final part of this chapter and learn how to make generative AI models execute actions with agents.

Building action models with agents

Agents are the newest feature in the generative AI world. They are powerful tools that enable the automation of tasks by allowing generative AI models to take actions on our behalf. They act as intermediaries between the generative AI models and external systems or services, facilitating the execution of tasks in the real world.

Under the hood, an agent "understands" what the user wants and calls a backend function that performs the action. The scope within which the agent can act is defined by an OpenAPI schema that it will use both to "understand" what it does and how to properly call the backend function.

So, in summary, to build an agent we need an OpenAPI schema, a backend function, and a knowledge base. The knowledge base is optional, but it can greatly improve a user's experience with the AI assistant.

For this section's exercise, we will build an agent that "knows" the available information about the AWS Competency program. A visual representation of the agent's application architecture is shown in *Figure 11.10*.

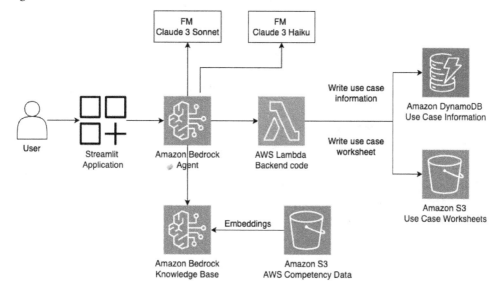

Figure 11.10 – Agent application architecture

This agent is going to build a simple worksheet with a use case's information, save the worksheet to Amazon S3, and register the information on a DynamoDB table for consultation. Let's get to it:

1. First, we need an OpenAPI schema defining the methods available for our agent. In this case, we will define two methods. The first one, `generateCaseSheet`, registers the use case information and builds the worksheet. The second, `checkCase`, takes the use case ID and returns information about it. As this is a long JSON file, we will not display it here. The complete code is available at `https://github.com/PacktPublishing/Bigdata-on-Kubernetes/tree/main/Chapter11/agent` folder. Copy this code and save it in an S3 bucket.

2. Next, we will define a Lambda function that will serve as a backend for the agent. The complete Python code for the function is available in the book's GitHub repository under the `Chapter 11/agent/function` folder. In your machine, create a folder named `function` and save this code as `lambda_function.py` in the `function` folder. This code defines a Lambda function that serves as the backend for a Bedrock agent. The function handles two different API paths: `/generateCaseSheet` and `/checkCase`. Let's go through the code block by block. After importing the necessary folders, we define two helper functions to extract parameter

values from the event object (`get_named_parameter` and `get_named_property`). The `generateCaseSheet` function is responsible for creating a new case sheet based on the provided information. It extracts the required parameters from the event object, generates a unique ID, creates a new Excel workbook using the `CaseTemplate` class, fills in the template with the provided parameters, saves the workbook to a temporary file, uploads it to an S3 bucket, and stores the case sheet information in a DynamoDB table. Finally, it returns a response object containing the case details. The `checkCase` function retrieves the case sheet information from the DynamoDB table based on the provided `caseSheetId` parameter and returns a response object containing the case details. The `lambda_handler` function is the entry point for the Lambda function. It determines the appropriate action based on the `apiPath` value in the event object. The function constructs the appropriate response object based on the action and returns it.

3. Next, inside the `function` folder, create a new file called `lambda_requirements.txt` where we will list the dependencies for the Lambda function code. In the `lambda_requirements.txt` file, type `openpyxl==3.0.10` and save it.

4. Now, before deploying the function, we need to create an IAM role that will give Lambda the necessary permissions. On the AWS console, go to the IAM page, choose **Roles** in the side menu, and click on **Create a new role**.

5. On the next page, select **AWS service** for the **Trusted entity type** and **Lambda** for the **Use case** (as shown in *Figure 11.11*). Click **Next**.

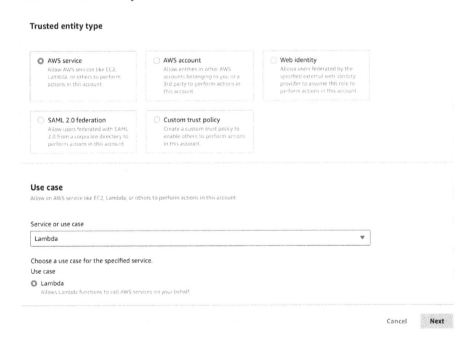

Figure 11.11 – Selecting trusted entity and AWS service

6. Now, we will select a permission policy. Choose **Administrator Access** and click **Next**. Remember that having such open permissions is *not* a good practice for production environments. You should set permissions only for the actions and resources needed.

7. Then, choose a name for your IAM role (`BDOK-Lambda-service-role`, for instance) and click on **Create role**.

8. Then, you will see the IAM **Roles** page again. Search for your created role and click on it (*Figure 11.12*).

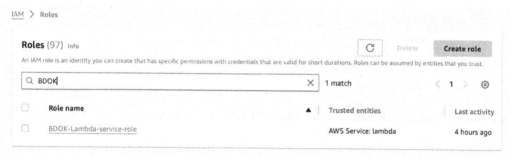

Figure 11.12 – Selecting your created IAM role

9. On the role's page, you will see the **Amazon Resource Name** (**ARN**) of the role. Copy it and save it for later. We will need that name to deploy the Lambda function.

10. Next, inside the `function` folder you created, create a new folder called `worksheet`. Copy two files from `https://github.com/PacktPublishing/Bigdata-on-Kubernetes/tree/main/Chapter11/agent/function/worksheet`, the first named `__init__.py` and the second named `template.py` and place those code files inside the `worksheet` folder. This code contains a class named `CaseTemplate` that builds an Excel worksheet with the `openpyxl` Python library.

11. Next, copy another two files in `https://github.com/PacktPublishing/Bigdata-on-Kubernetes/tree/main/Chapter11/agent/scripts` folder named `build_lambda_package.sh` and `create_lambda_function.sh`. Those files contain bash code that will install the dependencies for the Lambda function and deploy it to AWS.

12. Now, we will deploy our Lambda function. This is a good time to check whether your project structure is correct. The folder and file structure should look like this:

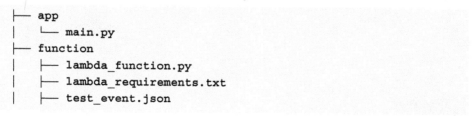

```
|       └── worksheet
|           ├── __init__.py
|           └── template.py
├── openapi_schema.json
└── scripts
    ├── build_lambda_package.sh
    └── create_lambda_function.sh
```

If your project structure is different, correct it to get this exact structure or the bash code will not work properly.

13. Now, go to the `scripts` folder and run the following commands:

```
sh build_lambda_package.sh
sh create_lambda_function.sh "<YOUR_ROLE_ARN>"
```

Remember to change <YOUR_ROLE_ARN> to the actual ARN of your Lambda IAM role. Now, we have some more work to do. Next, we will create the DynamoDB table to store information about the use cases.

Creating a DynamoDB table

DynamoDB is a fully managed NoSQL database service. It is a key-value and document database that can deliver single-digit millisecond performance at any scale. DynamoDB is optimized for running serverless applications and is designed to scale up or down automatically to meet demand, without having to provision or manage servers. It is particularly well suited for applications that need low-latency read and write access to data at any scale. Its extremely low latency makes it a very good choice for an AI assistant application. Let's get to it:

1. In the AWS console, navigate to the **DynamoDB** page. In the side menu, click on **Tables** and then, click on **Create table**.

2. On the next page, fill in **Table name** with `case-sheets` and the **Partition key** field with `caseSheetId`. Remember to select **Number** to indicate that this entry is a number, as shown in *Figure 11.13*. Leave all the other configurations to default and click **Create table**.

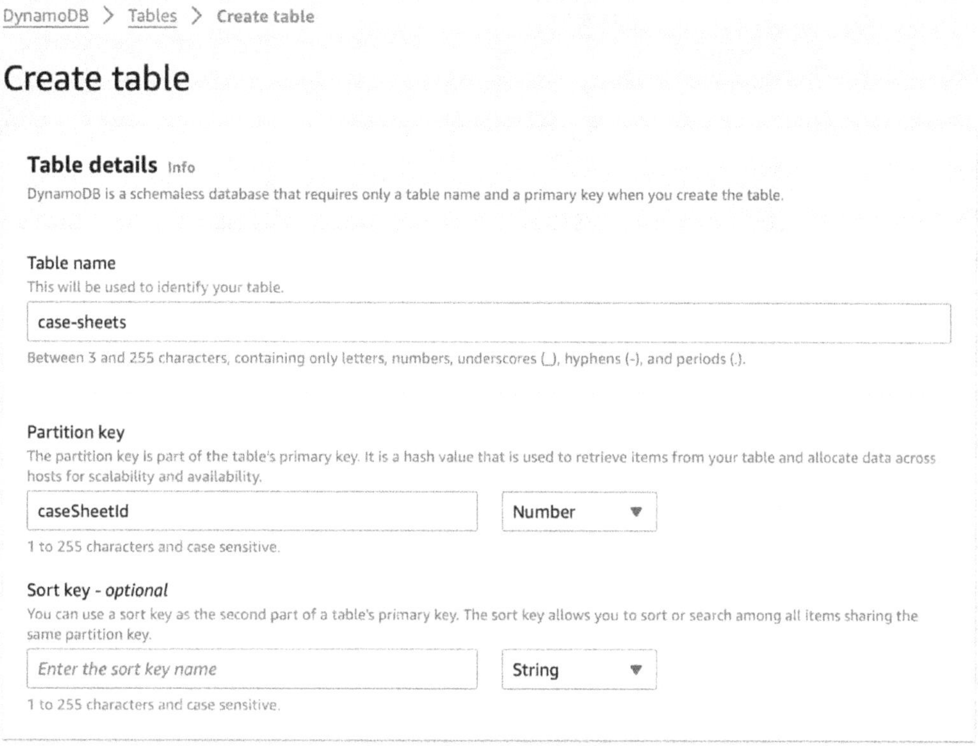

DynamoDB > Tables > Create table

Create table

Table details Info

DynamoDB is a schemaless database that requires only a table name and a primary key when you create the table.

Table name

This will be used to identify your table.

```
case-sheets
```

Between 3 and 255 characters, containing only letters, numbers, underscores (_), hyphens (-), and periods (.).

Partition key

The partition key is part of the table's primary key. It is a hash value that is used to retrieve items from your table and allocate data across hosts for scalability and availability.

```
caseSheetId
```
```
Number      ▼
```

1 to 255 characters and case sensitive.

Sort key - optional

You can use a sort key as the second part of a table's primary key. The sort key allows you to sort or search among all items sharing the same partition key.

```
Enter the sort key name
```
```
String      ▼
```

1 to 255 characters and case sensitive.

Figure 11.13 – Creating a DynamoDB table

In a few seconds, you should have your DynamoDB table ready for use. Now, we will configure the Bedrock agent.

Configuring the agent

Now, in the last part of this section, we will configure a Bedrock agent and link it to its backend Lambda function and the knowledge base database. Let's get to it:

1. First, in the AWS console, search for Bedrock and, in the side menu, click on **Agents**.

2. In the pop-up box, enter the name of your agent (aws-competency-agent) and click on **Create**.

3. Next, you will see the agent configuration page. Scroll down to **Select model** and choose the Anthropic model **Claude 3 Haiku** (you can also play with the other available models as you like).

4. In the **Instructions for the agent** field, set the initial instructions that will guide the foundational model in its behavior. You can type, for instance: You are a friendly AI assistant. Your main goal is to help AWS partner companies build case

sheets for the AWS Competency program, register those cases, and tell the user the information about the registered cases. When you generate a case sheet, always show back to the user the ID of the case sheet (id), the client's name (client), and the name of the case (casename) and confirm that the case was successfully created. Also, answer the questions of the user about what you can do and how you can help. This is a very important part of the agent configuration. Play with these instructions as much as you like. An example of this screen is shown in *Figure 11.14*.

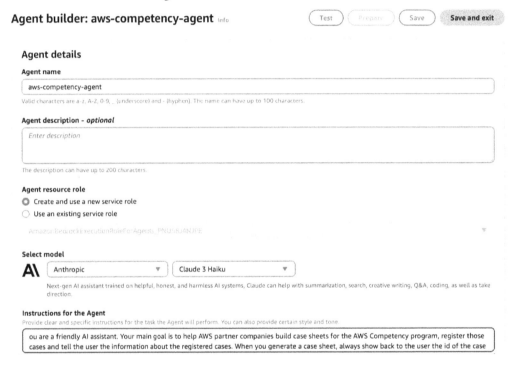

Figure 11.14 – Configuring the agent's instructions

5. After that, click on the **Save** button at the top of the page to make AWS create the necessary permission policy.

6. Next, scroll down to the **Action groups** section and click on **Add**.

7. On the next page, select a name for your action group. For **Action group type**, select **Define with API schemas**. In **Action group invocation**, select **Select an existing Lambda function** and select the Lambda function we just created (*Figure 11.15*).

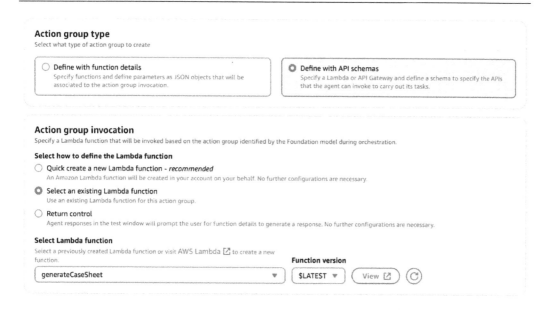

Figure 11.15 – Selecting a Lambda function for the agent's action group

8. Now, in the **Action group schema** section, choose **Select an existing API schema** and then, click on **Browse S3** to search for the OpenAPI schema we have saved on S3 (*Figure 11.16*). Then, click on **Create**.

Figure 11.16 – Selecting the OpenAPI schema

9. Next, in the **Knowledge base** section, click on **Add**.

10. Select the knowledge base we have created before and type some instructions for the agent on how to use it. For instance: This knowledge base contains information on the following AWS Competency programs: conversational AI, data and analytics, DevOps, education, energy, financial services, machine learning, and security. Make sure **Knowledge base status** is set to **Enabled** (*Figure 11.17*). Click **Save and exit**.

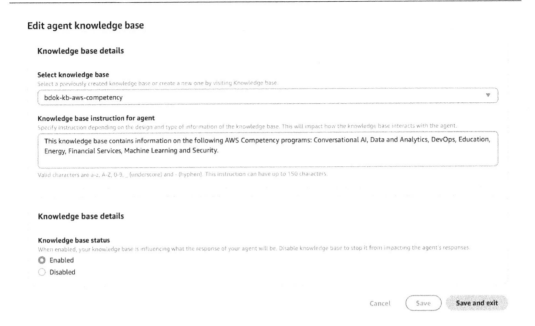

Figure 11.17 – Attaching a knowledge base to the agent

11. Now, you are back to the agent's editing page. Nothing else is needed here, so you can click on **Prepare** at the top to get your agent ready to run, and then click on **Save and exit**.

12. Now, you will be led back to the agent's main page. Scroll down to the **Aliases** section and click on **Create**.

13. Type in an alias name (aws-competency, for instance) and click on **Create Alias**.

Figure 11.18 – Creating an alias

14. Now, the last thing to do is register a permission on Lambda for this agent to trigger the function execution. On the agent's main page, copy the agent ARN.

15. Next, go to the **Lambda** page and click on the function we created for this exercise. On the function's main page, scroll down, click on **Configuration**, and then click on **Permissions** in the side menu (*Figure 11.19*).

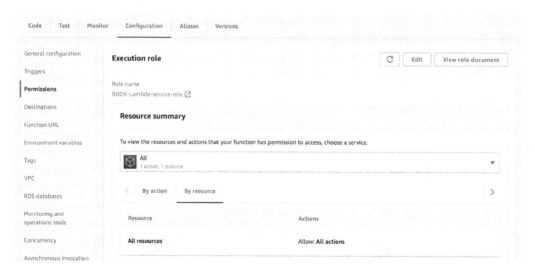

Figure 11.19 – Lambda permissions

16. Scroll down again to the **Resource-based policy statements** section and click on **Add permissions**.

17. On the next page, fill in the **Statement ID** box. For **Principal**, paste the agent ARN you copied from the Bedrock agent. For **Action**, choose `lambda:InvokeFunction` (*Figure 11.20*). Then, click on **Save**.

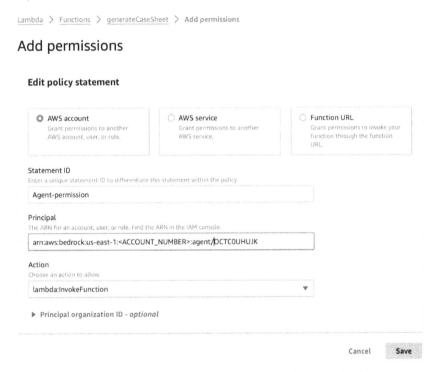

Figure 11.20 – Configuring the Lambda permission for the Bedrock agent

That's all the configuration we need to get the agent running. Now, it's time for deployment. Let's get our Streamlit application to Kubernetes.

Deploying the application on Kubernetes

Deploying the agent Streamlit application on Kubernetes follows the same path we did for the other two applications deployed before. The only thing different is that we must create a new `configmap` with the agent's ID and its alias ID:

1. Go to the **Agent** page in the AWS console and copy the agent's ID (in the top section) and the alias ID (in the bottom section).

2. Now, create `configmap` with those parameters with the following command:

```
kubectl create configmap agent-config --from-literal=agent_
alias_id=<YOUR_ALIAS_ID> --from-literal=agent_id=<YOUR_AGENT_ID>
-n genai
```

Remember to replace the <YOUR_ALIAS_ID> and <YOUR_AGENT_ID> placeholders with the actual values.

3. Build a custom image if you want to customize the application. If you don't, use the ready image from DockerHub (https://hub.docker.com/r/neylsoncrepalde/chat-with-claude-agent).

4. Next, we will define a deploy_agent.yaml file for the application and service deployment. The content for this file is available at https://github.com/PacktPublishing/Bigdata-on-Kubernetes/tree/main/Chapter11/agent folder.

5. With this file copied locally, now run the following:

```
kubectl apply -f deploy_agent.yaml -n genai
```

6. Wait for a few seconds for LoadBalancer to get started. Then, run the following to get the URL of LoadBalancer:

```
kubectl get svc -n genai
```

Paste it in a browser adding the correct port (:8501) to see the magic happening (*Figure 11.21*).

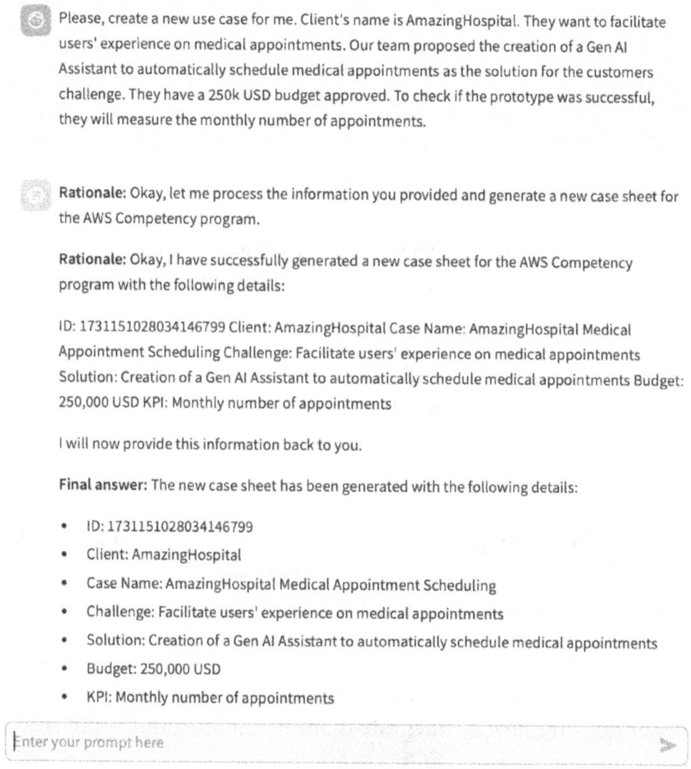

Figure 11.21 – AWS Competency agent application UI

Try inserting a prompt for the creation of a new use case as in *Figure 11.18*. Also, you can check specific information about this case passing the case's ID (*Figure 11.22*).

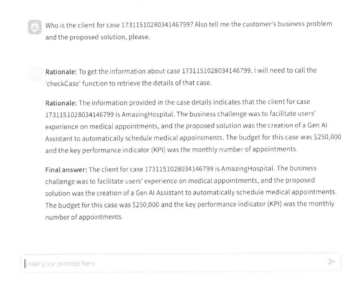

Chat with Bedrock Agent

AWS Competency Program

Who is the client for case 17311510280341467997 Also tell me the customer's business problem and the proposed solution, please.

Rationale: To get the information about case 17311510280341146799, I will need to call the 'checkCase' function to retrieve the details of that case.

Rationale: The information provided in the case details indicates that the client for case 17311510280341146799 is AmazingHospital. The business challenge was to facilitate users' experience on medical appointments, and the proposed solution was the creation of a Gen AI Assistant to automatically schedule medical appointments. The budget for this case was $250,000 and the key performance indicator (KPI) was the monthly number of appointments.

Final answer: The client for case 17311510280341146799 is AmazingHospital. The business challenge was to facilitate users' experience on medical appointments, and the proposed solution was the creation of a Gen AI Assistant to automatically schedule medical appointments. The budget for this case was $250,000 and the key performance indicator (KPI) was the monthly number of appointments.

Enter your prompt here

Figure 11.22 – Checking use case information with the agent

Play a little bit. Ask questions about the Competency program and try registering different cases. Also, you can check AWS DynamoDB and see the information ingested in our created table and check S3 to see the Excel files the agent created.

That is it! Congratulations! You have just deployed a full generative AI agent application that can perform tasks for you using natural language on Kubernetes.

Summary

In this chapter, we explored the exciting world of generative AI and learned how to harness its power on Kubernetes. We started by understanding the fundamental concepts of generative AI, its underlying mechanisms, and how it differs from traditional machine learning approaches.

We then leveraged Amazon Bedrock, a comprehensive suite of services, to build and deploy generative AI applications. We learned how to work with Bedrock's foundational models, such as Claude 3 Haiku and Claude 3 Sonnet, and how to integrate them into a Streamlit application for interactive user experiences.

Next, we delved into the concept of RAG, which combines the power of generative AI with external knowledge bases. We built a RAG system using Knowledge Bases for Amazon Bedrock, enabling our application to access and leverage vast amounts of structured data, improving the accuracy and relevance of the generated output.

Finally, we explored Agents for Amazon Bedrock, a powerful feature that allows generative AI models to automate tasks and take actions on our behalf. We learned how to build an agent, define its capabilities through an OpenAPI schema, and create the underlying Lambda function that serves as the backend for our agent.

Throughout this chapter, we gained hands-on experience in building and deploying generative AI applications on Kubernetes. The skills and knowledge acquired in this chapter are invaluable in today's rapidly evolving technological landscape. Generative AI is transforming industries and revolutionizing how we interact with and leverage AI. By mastering the tools and techniques presented in this chapter, you will be well equipped to build innovative and intelligent applications that can generate human-like content, leverage external knowledge sources, and automate tasks.

In the next chapter, we will discuss some important points needed for a production-ready Kubernetes environment, which we did not have space to discuss throughout the book.

12

Where to Go from Here

Congratulations on making it this far in your journey toward mastering big data on Kubernetes! By now, you've gained a solid understanding of the core concepts and technologies involved in running big data workloads on Kubernetes. However, as with any complex system, there's always more to learn and explore.

In this chapter, we'll guide you through the next steps in your development journey, covering some of the most important topics and technologies you should focus on as you move toward production-ready big data deployments on Kubernetes. We'll discuss crucial aspects such as **monitoring**, **service meshes**, **security**, **automated scalability**, **GitOps**, and **continuous integration/continuous deployment (CI/ CD)** for Kubernetes.

For each topic, we'll provide you with an overview and introduce you to the relevant technologies and tools, giving you a solid foundation to build upon. This will enable you to dive deeper into the areas that are most relevant to your specific use cases and requirements.

By the end of this chapter, you'll have a clear understanding of the key concepts and technologies that are essential for running big data workloads on Kubernetes in a production environment. You'll be equipped with the knowledge to make informed decisions about which tools and approaches to adopt, and you'll have a roadmap for further exploration and learning.

We'll also touch on the skills and team structure that will optimize your organization for success with this architecture.

Whether you're a seasoned Kubernetes user or just starting your journey, this chapter will provide you with valuable insights and guidance to take your big data on Kubernetes implementation to the next level.

In this chapter, we're going to cover the following main topics:

- Important topics for big data in Kubernetes
- What about team skills?

Important topics for big data in Kubernetes

As we approach the end of this book, it's important to recognize that the journey toward mastering big data on Kubernetes is far from over. Throughout the chapters, we've covered a wide range of topics, from deploying and managing big data applications on Kubernetes to optimizing performance and scalability. However, there are several crucial areas that we haven't had the opportunity to explore in depth, yet they are essential for building a robust and production-ready big data infrastructure on Kubernetes.

In this section, we'll take a closer look at some of the most important topics that you should familiarize yourself with as you continue your journey with big data and Kubernetes. These topics include Kubernetes monitoring and application monitoring, building a service mesh, security considerations, automated scalability, GitOps, CI/CD for Kubernetes, and Kubernetes cost control. While we won't delve into the intricate details of each topic, we'll provide an overview of the main concepts involved, the primary technologies used to implement these solutions on Kubernetes, and the biggest challenges you may encounter along the way.

Kubernetes monitoring and application monitoring

Monitoring is a critical aspect of any production-ready system, and it becomes even more crucial when dealing with complex big data applications running on Kubernetes. Kubernetes monitoring involves tracking the health and performance of the Kubernetes cluster itself, including the control plane, worker nodes, and various Kubernetes components. On the other hand, application monitoring focuses on applications running within the Kubernetes cluster, monitoring their performance, resource utilization, and overall health.

For Kubernetes monitoring, popular tools such as **Prometheus** and **Grafana** can be employed. These tools collect metrics from various Kubernetes components and provide visualizations and alerting mechanisms to help you stay on top of your cluster's health. Application monitoring, on the other hand, often relies on application-specific monitoring solutions or integrations with tools such as Prometheus or other third-party monitoring solutions such as **Datadog**, **Splunk**, and **Dynatrace**, for instance.

One of the biggest challenges in monitoring big data applications on Kubernetes is the sheer volume of data and the complexity of the applications themselves. Big data applications often consist of multiple components, each with its own set of metrics and monitoring requirements. Additionally, the distributed nature of these applications can make it challenging to correlate metrics across different components and gain a comprehensive understanding of the system's overall health.

Building a service mesh

As your big data applications on Kubernetes grow in complexity, managing network communication, observability, and traffic control can become increasingly challenging. This is where a service mesh comes into play. A service mesh is an infrastructure layer that sits between the application components and the underlying network, providing a consistent and centralized way to manage service-to-service communication, traffic routing, and observability.

Popular service mesh solutions for Kubernetes include **Istio**, **Linkerd**, and **Consul**. These tools provide features such as load balancing, circuit breaking, retries, and traffic routing, as well as observability capabilities such as distributed tracing and metrics collection. By implementing a service mesh, you can decouple these cross-cutting concerns from your application code, making it easier to manage and maintain your big data applications.

However, introducing a service mesh into your Kubernetes environment also comes with its own set of challenges. Service meshes can add complexity and overhead to your system, and their configuration and management can be non-trivial. Additionally, ensuring the compatibility of your applications with the service mesh and understanding the performance implications are crucial considerations.

Security considerations

Security is a paramount concern when working with big data applications on Kubernetes, as these systems often deal with sensitive data and must comply with various regulatory requirements. Kubernetes provides several built-in security features, such as **role-based access control** (**RBAC**), network policies, and secrets management. However, implementing a comprehensive security strategy requires a multilayered approach that addresses various aspects of your big data infrastructure.

Some key security considerations include securing the Kubernetes control plane and worker nodes, implementing network segmentation and isolation, managing secrets and sensitive data, and ensuring compliance with industry standards and regulations. Tools such as **Falco**, **Kubesec**, and **Kube-bench** can help you assess and enforce security best practices within your Kubernetes environment.

One of the biggest challenges in securing big data applications on Kubernetes is the complexity and distributed nature of these systems. Big data applications often consist of multiple components, each with its own set of security requirements and potential vulnerabilities. Additionally, ensuring the secure handling and storage of large volumes of data, while maintaining performance and scalability, can be a significant challenge.

Automated scalability

One of the key benefits of running big data applications on Kubernetes is the ability to scale resources dynamically based on demand. However, achieving effective and efficient automated scalability requires careful planning and implementation. Kubernetes provides built-in mechanisms for horizontal and vertical scaling, such as the **Horizontal Pod Autoscaler** (**HPA**) and the **Vertical Pod Autoscaler** (**VPA**). These tools allow you to automatically scale the number of replicas or adjust resource requests and limits based on predefined metrics and thresholds.

In addition to the built-in Kubernetes scaling mechanisms, there are third-party solutions such as **Kubernetes Event-driven Autoscaling** (**KEDA**) that can provide more advanced scaling capabilities.

KEDA is an open source Kubernetes scaling solution that allows you to automatically scale your workloads based on event-driven patterns. It provides a simple and lightweight way to define event sources and scale your deployments based on the number of events that need to be processed.

However, implementing effective automated scalability for big data applications on Kubernetes can be challenging. Big data applications often have complex resource requirements and dependencies, making it difficult to define appropriate scaling thresholds and metrics. Additionally, ensuring the seamless scaling of stateful components, such as databases or message queues, can introduce additional complexities.

GitOps and CI/CD for Kubernetes

As your big data infrastructure on Kubernetes grows in complexity, managing and deploying changes becomes increasingly challenging. This is where GitOps and CI/CD practices come into play. GitOps is a methodology that treats infrastructure as code, using Git as the **single source of truth** (**SSOT**) for declarative infrastructure definitions. CI/CD, on the other hand, is a set of practices and tools that enable the automated building, testing, and deployment of applications.

Popular GitOps tools for Kubernetes include **Argo CD**, **Flux**, and **Jenkins X**, while CI/CD solutions such as **Jenkins**, **GitLab CI/CD**, and **GitHub Actions** can be integrated with Kubernetes to enable automated deployments. By adopting GitOps and CI/CD practices, you can streamline the deployment process, ensure consistency across environments, and reduce the risk of human errors.

One of the biggest challenges in implementing GitOps and CI/CD for big data applications on Kubernetes is the complexity of the applications themselves. Big data applications often consist of multiple components with intricate dependencies and configurations. Ensuring the correct order of deployments, handling stateful components, and managing complex configurations can be a significant hurdle. Additionally, integrating GitOps and CI/CD practices with existing infrastructure and processes can require significant effort and cultural shifts within your organization.

Kubernetes cost control

Kubernetes cost control is a critical aspect of managing and optimizing resources and expenses associated with running applications and workloads on a Kubernetes cluster. As organizations adopt Kubernetes for their application deployments, they often face challenges in understanding and controlling the costs associated with the underlying infrastructure, such as **virtual machines** (**VMs**), storage, and networking resources.

Cost control in the context of Kubernetes involves monitoring, analyzing, and optimizing resource utilization and spending across the entire Kubernetes ecosystem. It helps organizations gain visibility into their cloud expenditure, identify areas of inefficiency or over-provisioning, and implement strategies to reduce costs without compromising application performance or availability.

The importance of cost control in Kubernetes stems from the dynamic and scalable nature of containerized applications. Kubernetes enables automatic scaling of resources based on demand, which can lead to unexpected cost increases if not properly managed. Additionally, organizations may inadvertently over-provision resources or leave unused resources running, resulting in unnecessary expenses.

Several tools and solutions have emerged to address the need for cost control in Kubernetes environments. One of the most popular open source tools is Kubecost, which provides real-time cost monitoring, allocation, and optimization for Kubernetes clusters. Kubecost integrates with various cloud providers, such as **Amazon Web Services (AWS)**, **Azure,** and **Google Cloud Platform (GCP)**, and offers features such as cost allocation by namespace, deployment, or service, cost forecasting, and recommendations for cost optimization.

Kubecost works by collecting metrics from the Kubernetes API and cloud provider APIs, analyzing resource utilization and pricing data, and presenting cost information through a user-friendly interface or via integrations with other monitoring and alerting tools. It allows teams to identify cost drivers, set budgets, and receive alerts when costs exceed predefined thresholds.

One of the main challenges is the complexity of Kubernetes itself, with its numerous components and configurations that can impact resource utilization and costs. Additionally, organizations may struggle with accurately attributing costs to specific applications or teams, especially in multi-tenant environments. Another challenge lies in striking the right balance between cost optimization and application performance. Over-aggressive cost-saving measures, such as under-provisioning resources or disabling auto-scaling, can lead to performance degradation or application downtime, which can ultimately result in lost revenue or customer dissatisfaction.

To effectively address cost control in Kubernetes, organizations must adopt a holistic approach that involves continuous monitoring, analysis, and optimization. This includes implementing cost governance policies, setting budgets and alerts, regularly reviewing resource utilization and rightsizing resources, and fostering a culture of cost awareness across development, operations, and finance teams.

Running Kubernetes in a production environment poses a lot of technology challenges, but this is only half of the game. People skills, team building, and organizational shared knowledge are extremely important to have a successful implementation. Next, we are going to discuss the skills needed to have a production-ready environment to work with big data on Kubernetes and for technical team building.

What about team skills?

As we discussed in the previous section, implementing and managing big data applications on Kubernetes involves a wide range of concepts and technologies. To ensure the success of your big data initiatives on Kubernetes, it's crucial to have a team with the right skills and expertise. In this section, we'll explore the key skills required for each of the topics mentioned earlier and discuss how these skills can be mapped to specific roles within your technical team.

Key skills for monitoring

Effective monitoring requires a combination of skills from different domains. First and foremost, you'll need team members with a deep understanding of Kubernetes and its various components. They should be proficient in deploying and configuring monitoring tools such as Prometheus and Grafana, as well as integrating them with the Kubernetes ecosystem. Additionally, they should have experience in designing and implementing monitoring strategies, defining relevant metrics, and setting up alerting and notification systems.

For application monitoring, you'll need team members with expertise in the specific big data technologies and frameworks you're using. They should be able to identify and instrument critical components of your applications, understand their performance characteristics, and define appropriate monitoring metrics. Furthermore, they should be skilled in integrating application monitoring solutions with the overall monitoring infrastructure.

These skills are typically found in roles such as *site reliability engineers (SREs)*, *DevOps engineers*, and *cloud engineers*. These professionals often have a strong background in systems administration, automation, and monitoring, combined with experience in cloud computing and containerization technologies such as Kubernetes.

Building a service mesh

Implementing a service mesh requires a deep understanding of networking concepts, service-to-service communication patterns, and observability principles. Your team should have members skilled in deploying and configuring service mesh solutions such as Istio, Linkerd, or Consul. They should be proficient in defining traffic routing rules, implementing security policies, and leveraging observability features provided by the service mesh.

Additionally, they should have experience in integrating the service mesh with your existing applications and ensuring compatibility with the various components of your big data infrastructure. These skills are often found in roles such as *platform engineers*, *cloud engineers*, and *DevOps engineers*, with a strong focus on networking and observability.

Security considerations

Securing big data applications on Kubernetes requires a multidisciplinary approach, combining expertise in various security domains. Your team should have members skilled in implementing and managing Kubernetes security controls, such as RBAC, network policies, and secrets management. They should be proficient in hardening Kubernetes clusters, enforcing security best practices, and conducting regular security audits and vulnerability assessments.

Furthermore, you'll need team members with expertise in data security and compliance, particularly in the context of big data applications. They should be knowledgeable about industry standards and regulations, such as the **General Data Protection Regulation (GDPR)**, the **Health Insurance Portability and Accountability Act (HIPAA)**, or the **Payment Card Industry Data Security Standard (PCI DSS)**, and be able to implement appropriate security measures to ensure compliance.

These skills are typically found in roles such as *security engineers*, *cloud security analysts*, and *compliance specialists*. These professionals often have a strong background in cybersecurity, risk management, and regulatory compliance, combined with experience in cloud computing and containerization technologies.

Automated scalability

Implementing effective automated scalability for big data applications on Kubernetes requires a combination of skills in application architecture, performance optimization, and automation. Your team should have members skilled in designing and implementing scalable and resilient applications, understanding the resource requirements and scaling patterns of different components, and defining appropriate scaling metrics and thresholds.

They should also be proficient in using the aforementioned Kubernetes scaling mechanisms such as HPA, VPA, and KEDA. Additionally, they should have experience in automating scaling processes, integrating with monitoring and alerting systems, and ensuring the seamless scaling of stateful components.

These skills are often found in roles such as *cloud engineers*, *DevOps engineers*, and *SREs*. These professionals typically have a strong background in cloud computing, containerization, and automation, combined with experience in performance optimization and application architecture.

Skills for GitOps and CI/CD

Adopting GitOps and CI/CD practices for big data applications on Kubernetes requires a combination of skills in version control, **infrastructure as code (IaC)**, and automation. Your team should have members skilled in using Git and Git-based tools such as Argo CD, Flux, or Jenkins X for managing and deploying Kubernetes resources. They should be proficient in writing and maintaining declarative infrastructure definitions and have experience in implementing GitOps workflows and best practices.

Additionally, they should have expertise in setting up and configuring CI/CD pipelines, integrating them with Kubernetes, and automating build, testing, and deployment processes. They should be skilled in handling complex application dependencies, managing stateful components, and ensuring consistent deployments across different environments.

These skills are typically found in roles such as *DevOps engineers*, *platform engineers*, and *release engineers*. These professionals often have a strong background in software development, automation, and IaC, combined with experience in cloud computing and containerization technologies such as Kubernetes.

Cost control skills

Effective cost control in Kubernetes environments requires a collaborative effort from various team members with diverse skills and expertise. These professionals should possess a combination of technical knowledge, analytical abilities, and a deep understanding of the organization's business objectives and cost constraints.

One of the key roles in cost control is the *cloud cost engineer* or *FinOps engineer*. These professionals should have a strong grasp of cloud computing technologies, including Kubernetes, container orchestration, and cloud provider services. They should have a deep understanding of pricing models, resource utilization patterns, and cost optimization strategies.

Another crucial role is the *cloud architect* or *platform engineer*. These individuals should have extensive experience in designing and implementing cloud-native architectures, including Kubernetes clusters, microservices, and serverless functions. They should be adept at optimizing resource allocation, implementing auto-scaling strategies, and leveraging cost-effective cloud services. Their expertise in IaC and CI/CD pipelines is also essential for efficient resource management and cost control.

Developers and DevOps engineers are also critical contributors to cost control efforts. They should have a deep understanding of application architectures, resource requirements, and performance characteristics. Their ability to write efficient and optimized code, implement containerization best practices, and leverage auto-scaling and rightsizing techniques can significantly impact resource utilization and costs.

In recent years, a new role has emerged in organizations focused on cost control: the *cost champion*. This individual acts as an advocate for cost awareness and optimization across teams. Cost champions work closely with developers, operations, and finance teams to promote cost-conscious practices, provide training and guidance, and ensure that cost considerations are integrated into the **software development life cycle** (SDLC). They should have strong communication and leadership skills, as well as a deep understanding of the organization's cost structure and business objectives.

Effective cost control in Kubernetes requires a collaborative effort from cross-functional teams with diverse skills and expertise. By fostering a culture of cost awareness, leveraging the right tools and technologies, and empowering teams with the necessary knowledge and resources, organizations can achieve significant cost savings while maintaining application performance and availability.

It's important to note that while these roles and skill sets provide a general guideline, the specific requirements may vary depending on the size and complexity of your organization, as well as the specific big data technologies and frameworks you're using. In some cases, you may need to combine or distribute these responsibilities across multiple roles or teams.

By building a team with the right skills and expertise, and fostering a supportive and collaborative environment, you'll be well equipped to tackle the challenges of implementing and managing big data applications on Kubernetes and unlock the full potential of this powerful combination of technologies.

Summary

In this chapter, we explored the next steps in your journey toward mastering big data on Kubernetes. We covered several important topics that are essential for building a robust and production-ready big data infrastructure on Kubernetes, including Kubernetes monitoring and application monitoring, building a service mesh, important security considerations to take into account in a production environment, scalability automation methods, GitOps and CI/CD for Kubernetes, and Kubernetes cost control.

We also discussed the importance of having a team with the right skills and expertise to tackle these challenges successfully. We covered the key roles and skill sets required for each of the topics mentioned, including SREs, DevOps engineers, cloud engineers, security engineers, and release engineers.

Congratulations on reaching the end of this book! You've taken a significant step toward mastering the art of running big data workloads on Kubernetes. The journey ahead may be challenging, but the knowledge and skills you've acquired will serve as a solid foundation for your future endeavors.

Remember – the field of big data and Kubernetes is constantly evolving, with new technologies and best practices emerging regularly. Embrace a mindset of continuous learning and stay curious.

Don't be afraid to experiment and try new approaches. Kubernetes and big data offer a vast playground for innovation, and your unique perspective and experiences can contribute to finding customized solutions that meet the needs of your projects, your company, or your customers.

Finally, remember that success in this field is not a solo endeavor. Attend conferences, participate in online communities, and engage with experts in the field to stay up to date with the latest developments. Collaborate with your team, share your knowledge, and learn from others. Together, you can overcome challenges, solve complex problems, and push the boundaries of what's possible with big data on Kubernetes.

Congratulations once again, and best of luck on your exciting journey ahead!

Index

packtpub.com

Subscribe to our online digital library for full access to over 7,000 books and videos, as well as industry leading tools to help you plan your personal development and advance your career. For more information, please visit our website.

Why subscribe?

- Spend less time learning and more time coding with practical eBooks and Videos from over 4,000 industry professionals

- Improve your learning with Skill Plans built especially for you

- Get a free eBook or video every month

- Fully searchable for easy access to vital information

- Copy and paste, print, and bookmark content

Did you know that Packt offers eBook versions of every book published, with PDF and ePub files available? You can upgrade to the eBook version at packtpub.com and as a print book customer, you are entitled to a discount on the eBook copy. Get in touch with us at customercare@packtpub.com for more details.

At www.packtpub.com, you can also read a collection of free technical articles, sign up for a range of free newsletters, and receive exclusive discounts and offers on Packt books and eBooks.

Other Books You May Enjoy

If you enjoyed this book, you may be interested in these other books by Packt:

Data Engineering with Databricks Cookbook

Pulkit Chadha

ISBN: 978-1-83763-335-7

- Perform data loading, ingestion, and processing with Apache Spark
- Discover data transformation techniques and custom user-defined functions (UDFs) in Apache Spark
- Manage and optimize Delta tables with Apache Spark and Delta Lake APIs
- Use Spark Structured Streaming for real-time data processing
- Optimize Apache Spark application and Delta table query performance
- Implement DataOps and DevOps practices on Databricks
- Orchestrate data pipelines with Delta Live Tables and Databricks Workflows
- Implement data governance policies with Unity Catalog

Data Engineering with Scala and Spark

Eric Tome, Rupam Bhattacharjee, David Radford

ISBN: 978-1-80461-258-3

- Set up your development environment to build pipelines in Scala
- Get to grips with polymorphic functions, type parameterization, and Scala implicits
- Use Spark DataFrames, Datasets, and Spark SQL with Scala
- Read and write data to object stores
- Profile and clean your data using Deequ
- Performance tune your data pipelines using Scala

Packt is searching for authors like you

If you're interested in becoming an author for Packt, please visit `authors.packtpub.com` and apply today. We have worked with thousands of developers and tech professionals, just like you, to help them share their insight with the global tech community. You can make a general application, apply for a specific hot topic that we are recruiting an author for, or submit your own idea.

Share Your Thoughts

Now you've finished *Big Data on Kubernetes*, we'd love to hear your thoughts! Scan the QR code below to go straight to the Amazon review page for this book and share your feedback or leave a review on the site that you purchased it from.

`https://packt.link/r/1-835-46214-6`

Your review is important to us and the tech community and will help us make sure we're delivering excellent quality content.

Download a free PDF copy of this book

Thanks for purchasing this book!

Do you like to read on the go but are unable to carry your print books everywhere?

Is your eBook purchase not compatible with the device of your choice?

Don't worry, now with every Packt book you get a DRM-free PDF version of that book at no cost.

Read anywhere, any place, on any device. Search, copy, and paste code from your favorite technical books directly into your application.

The perks don't stop there, you can get exclusive access to discounts, newsletters, and great free content in your inbox daily

Follow these simple steps to get the benefits:

1. Scan the QR code or visit the link below

https://packt.link/free-ebook/978-1-83546-214-0

2. Submit your proof of purchase
3. That's it! We'll send your free PDF and other benefits to your email directly